ENVIRONMENT AND URBANIZATION IN MODERN ITALY

HISTORY OF THE URBAN ENVIRONMENT

Martin V. Melosi and Joel A. Tarr, Editors

ENVIRONMENT AND URBANIZATION IN MODERN ITALY

FEDERICO PAOLINI

UNIVERSITY OF PITTSBURGH PRESS

Published by the University of Pittsburgh Press, Pittsburgh, Pa., 15260

Manufactured in the United States of America
Printed on acid-free paper
10 9 8 7 6 5 4 3 2 1

Cataloging-in-Publication data is available from the Library of Congress

ISBN 13: 978-0-8229-4593-2
ISBN 10: 0-8229-4593-2

Cover photo: *Acqua alta floods in Piazza San Marco*, Wolfgang Moroder.
Cover design: Joel W. Coggins

CONTENTS

ACRONYMS & ABBREVIATIONS

ACI	Automobile Club d'Italia (Automobile Club of Italy)
ANAS	Azienda Nazionale Autonoma delle Strade Statali (National Road Board)
ANCI	Associazione Nazionale Comuni Italiani (National Association of Italian Districts)
ARCI	Associazione Ricreativa e Culturale Italiana (Italian Recreation and Cultural Association)
ARPAT	Agenzia Regionale per la Protezione Ambientale della Toscana (Tuscany Regional Agency for Environmental Protection)
CENSIS	Centro Studi Investimenti Sociali (Center of Studies for Social Investments)
EC	European Commission
EEC	European Economic Community
EIA	Energy Information Administration
ENEA	Comitato Nazionale per la Ricerca e lo Sviluppo dell'Energia Nucleare e delle Energie Alternative (Italian National Agency for New Technologies, Energy, and Sustainable Economic Development)
ENI	Ente Nazionale Idrocarburi (National Hydrocarbon Corporation)
ENPA	Ente Nazionale Protezione Animali (National Association for the Protection of Animals)
EPIA	European Photovoltaic Industry Association
FAI	Fondo per l'Ambiente Italiano (Italian Environmental Fund)
FIGC	Federazione Giovanile Comunista Italiana (Italian Communist Youth Federation)
GRE	Gruppi di Ricerca Ecologica (Ecological Research Groups)
GWh	gigawatt hours
IEA	International Energy Agency
INU	Istituto Nazionale di Urbanistica (National Town Planning Institute)
IRPET	Istituto Regionale Programmazione Economica della Toscana (Tuscany Regional Institute for Economic Planning)
ISTAT	Istituto Nazionale di Statistica (National Institute of Statistics)
kW	kilowatt
kWh	kilowatt hour
kW/h	kilowatts per hour

LAC	Lega Anticaccia (Antihunting League)
LAV	Lega anti Vivisezione (Antivivisection League)
LIPU	Lega Italiana Protezione Uccelli (Italian League for the Protection of Birds)
LNG	liquid natural gas
LULU	Locally Unwanted Land Uses
Mtoe	million tons oil equivalent
MW	megawatt
NGO	Organizzazione non governativa (Non-governmental Organization)
OECD	Organizzazione per la cooperazione e lo sviluppo economico (Organization for Economic Cooperation and Development)
PIF	Piano Intercomunale Fiorentino (Urban Development Plan for the Metropolitan Area of Florence)
pkm	passenger-kilometer
PM_{10}	fine particulate matter
tkm	tonne-kilometer
toe	tons oil equivalent
TWh	terawatt hours
VAS	Verdi Ambiente e Società (Greens, Environment and Society)
WHO	Organizzazione Mondiale della Sanità (World Health Organization)
Wp	Watt peak
WWF	Fondo Mondiale per la Natura (World Wildlife Fund)

ENVIRONMENT AND URBANIZATION IN MODERN ITALY

INTRODUCTION

From the second half of the 1940s, when postwar reconstruction began in Italy, there were essentially three driving forces of environmental change. The first was apparent in the uncontrollable process of urban sprawl, fueled by considerable migratory flows from the countryside and southern regions toward the cities where large-scale productive activities were beginning to amass. The main consequence of demographic growth was an urban expansion model—following an explosion in building favored by a cartel of interests that united builders, real estate professionals, landowners, and the investment banks—characterized by the overseeing of scheduled settlements, by an infrastructural system tilted toward road networks, and by the lack of public services (e.g., greenspaces, public facilities). This model was repeated in all the cities that were spreading like wildfire, creating new urban areas built in a disorderly manner on portions of land set among industrial premises and inhabited chiefly by factory workers and their families.[1]

The second driving force of environmental change was unruly industrial development, which was tolerated since it was seen as the necessary tribute to be paid to progress and modernization. The paradigm of modernization was understood as the "path of exit from the rural condition, experienced as a condition of material scarcity in a static, restricted and introverted social and cultural context."[2] Precisely this paradigm rendered possible a system of industrial relations that hinged on a social contract in which enterprises were allowed to liberally exploit natural resources in exchange for supplying employment, the economic resources indispensable to freeing oneself from poverty. The signatories of the social contract (enterprises, workers, citizens, policy makers, institutions) did not at all consider any possible negative consequences of such a development model, which is to say the *degenerative metabolism* by which industries and urban centers appropriated resources in their entirety (waterways, sources of energy, raw materials, the land) and returned them in the form of degraded *metabolites* (biologically dead stretches of waters, polluted soil, air tainted by the emissions from houses, factories, and the means of transportation). The endurance of the social contract was made substantially possible by collective repression of the environmental damages, which were also clearly visible and, above all, continually denounced by scientists. Repression of environmental problems, then, was not fed by scarce collective consciousness or the absence of information, but rather by awareness that the benefits produced by this development model—in the eyes of the great majority of citizens and of political and economic policy makers—far outweighed the disadvantages.[3]

The second half of the 1960s saw the first manifestations of a third powerful driving force of environmental change: mass consumption. Failure to perceive that individual use of the environment generates a collective damage lay at the heart of the proliferation of the most common styles of consumption, such as the use of motor vehicles (which emit harmful substances that contribute to pollution of urban air and are partially responsible for the greenhouse effect) and the purchase of prepackaged goods (which contribute to the exponential increase in solid waste, whose disposal may generate metabolites dangerous to human health, such as dioxins).[4] Not all collective models of consumption are of a hedonistic nature—take the example of chemical products that can improve harvest yields or eliminate garden-infesting weeds—but they have played a role of fundamental importance in the onset of environmental crises, especially in urban areas.[5] Industrial production has in fact the aim of satisfying individual and collective needs; in market economy societies the quantity and quality of goods to be produced are determined by consumers' choices which, for all that they are influenced by sociocultural and psychological motivations (fueled, for example, by advertising), are chiefly induced by motives of a utilitarian nature. The examples are numerous: the car (whose success was essentially determined by improvement in the freedom of personal movement, not comparable to any means of public transportation); the washing machine (whose fortune was due to the enormous consumer buy-in, freeing them from the slavery of washing by hand); prepackaged food products (which meant quickly prepared meals at limited costs), and also plastic materials, which made it possible to produce objects that rapidly became widely used.

As Jared Diamond has efficiently shown, the history of humanity offers numerous examples of environmental crises caused by collective behaviors and lifestyles of societies that were unable to measure the impact of their actions on the environment.[6] On a par with the Danes in Greenland and the inhabitants of Easter Island, the postwar society repressed the environmental damage it was causing because it evidently saw the advantages produced by the development model as being far greater than the disadvantage of living in an ecosystem deeply modified by anthropic activities responsible for a growing environmental deterioration. Working in the factory, traveling by car, buying in supermarkets, living in newly built housing outside the old town centers, taking part in the rituals of consumer society (e.g., Saturday afternoon shopping, weekend trips, summer holidays at the seaside or in the mountains) were activities with a higher social value in comparison to the possibility of living in a more salubrious and less degraded ecosystem.

In this situation, up to the mid-1970s the necessary conditions for tackling environmental problems with serious protectionist and rehabilitation interventions were lacking. The whole country was paralyzed by the paradigm of modernization: the rhetorical imperatives were to modernize infrastructures and lifestyles in order to bring Italy into the assembly of industrialized and developed nations. Collective adherence to the development model was almost total. Generalized consent contributed to repression of the damages that economic growth and urbanization were inflicting on the environment; the material advantages were so evident and coveted that environmental damages were considered a modest tribute to pay for progress.

The context began to change, very slowly, only around the mid-1970s when the environmental crisis became evident and the need to supply a remedy could no longer be put off. This was made possible by the concurrence of the first industrial restructuring, following a slowing down of the economy caused by the oil crisis of 1973–1974. The first companies that closed down or relocated factories built in the new suburbs—or, attracted by favorable financing, opened new premises far from the big central-northern cities, thus emptying great portions of urban areas—were an early sign of the clear transformations brought about by a development model based on the twofold term *industrialization/urbanization*. Those who were left unemployed, residents who found themselves obliged to live next to the now lifeless industrial pachyderms, also began to become aware of the environmental damages produced by development. The first rehabilitation policies were set in motion at both national and local levels at this time. The interventions, however, were rarely incisive and continued to be vigorously obstructed by pressure groups, such as the unions of industrialists. We need only consider, for example, the events of Law No. 319 on waterways: the law was first emptied of content and then the application of its regulations was continuously put off until, almost twenty years later, the European Commission (EC) forced Italy to get in line with its directives.[7]

That said, the knotty question of the attitudes of workers and trade unions must be faced. It is entirely too easy—and demagogic—to place the blame for delayed implementation of environmental policies exclusively on the bosses, which, far more frequently than one might think, coincided with those of the workers. Companies certainly tried in every way (licit and illicit) to delay application of the (few) regulations that existed, and very likely also concealed the danger of certain work processes from the workers. But for decades the workers themselves had subordinated the safeguarding of their health and the environment to the guarantee of a job; this is demonstrated by the people of Prato, who were proud to see the waters of the Bisenzio River change color because those unnatural shades meant work and well-being. But there were also quarrels between citizens who protested about the smells and fumes produced by the factories and the workers who feared losing their occupations because of those protests.

The first real break came in the second half of the 1980s, simultaneously with the creation of the Ministry of the Environment and the victory of environmental petitions in the 1987 referendum on the civil use of nuclear power. These first significant changes in the attitude with which environmental problems were tackled gradually became increasingly incisive in the course of the 1990s, for two reasons: First, the great numbers of adherents achieved by the environmental movement after the accident at Chernobyl contributed to the subsequent electoral results obtained by the Green Party, which gained the Ministry of the Environment in center-left coalitions. Second, European treaties, by constructing a more solid and efficacious corpus of legislation, obliged reluctant countries such as Italy to produce regulations in line with European directives. In brief, the hetero-management of the EC played a role of fundamental importance that forced the European Union (EU) signatories—even the recalcitrant Mediterranean countries—to take care of their environmental problems within an ecosystem framework they had previously ignored.

There was a third cause, resting outside the purview of political and institutional decision makers: the rapid deindustrialization already begun during the late 1980s and then accelerated in the course of the 1990s. Progressive industrial desertification favored operations of environmental rehabilitation and made application of the regulations much easier. In short, a significant part of environmental improvements should be ascribed not to environmental policies but to the fact that the factories were moved elsewhere, exporting to other places the environmental problems that had previously afflicted Italian urban areas.

In this context the book (which brings together texts written from 2004 to 2016) seeks to delineate a path of research that lies within the vein of relationships between urban areas and the environment.[8] *Environment and Urbanization in Modern Italy* concentrates on the twentieth century and particularly on its last five decades, when the intensification of urbanization and industrialization led to a massive increase in the exploitation of natural resources.

PART I

NATURAL RESOURCES AND THEIR USES

There are two primary reasons for studying certain areas of Tuscany (a region in central Italy). The first concerns the urban development prerogatives of a region which was rapidly transforming from countryside, villages, and small to medium-size towns into a highly industrial area immersed in a profoundly urbanized rural context. The second regards the nature of the industrial development model, characterized by a clear predominance of small to medium-size businesses. It is precisely this aspect which makes study of the Tuscan case interesting because it allows us to verify how the advent of the environmental crisis is not traceable simply to the presence of great industrial complexes but to a more complex series of factors.

Investigating a case such as Tuscany, one is dissuaded from falling back on interpretations that tend to emphasize the role of industry and consequently explain the environmental crisis as a result of multiform interactions between urban *metabolisms* (flows of materials that enter cities and emerge as manufactured products or waste), exploited natural resources, and cultural attitudes (styles of consumption, residents' relationship with the natural environment).[1] Because it stands outside the geographical perimeter comprising the Genoa-Milan-Turin industrial triangle and the Po Valley as far as Bologna, the hub of urban and industrial development over the second half of the twentieth century, Tuscany is an incisive case study for investigating the crisis of urban government in the decades of accelerated development: that combination of dynamics which Martin Melosi has aptly defined as the "urban crisis in the age of ecology."[2]

Tuscany, a link between northern and southern Italy, embodies the characteristics of the country's history of urban environment, precisely as they have been categorized by Simone Neri Serneri and Gabriella Corona:[3] a relatively high population density distributed very differently among plains and hilly areas (the former much less extensive but far more populous); rapid and intense development, albeit in an economic framework featuring a prevalence of small and medium-size businesses; the presence of ongoing conflict between local government planning policies and interest groups intent on defending their specific advantages; a rapid acceleration in urbanization within a picture distinguished by widespread wildcat building in urban areas, where residential property shares space with industry; intense conflict between towns and those rural areas whose natural resources (water first and foremost) have been exploited to the exclusive advantage of urban areas.

The case of Tuscany is also interesting because it confirms another reflection common in Italian historiography, namely that the environmental question, as Salvatore Adorno and Simone Neri Serneri have written,[4] emerged only in the course of the 1970s, when the industrialization model established in the aftermath of the Second World War had fallen deeply into crisis. As in the rest of Italy—characterized by mature-technology industrial districts consisting of hundreds of medium-size and small businesses (often very small)—the environmental question in Tuscany emerged between the late 1970s and early 1980s, and its result was a *paper environmentalism*, because environmental policies were chiefly a rhetorical device and little else, as much at the national as the local level. While institutions and political forces made a great display of environmental rhetoric on the communications front, at the formal level their actions sought to delay the application of legislative regulations capable of kick-starting environmental rehabilitation. Having long studied the Tuscan case,[5] one has a strong sensation that institutional vagueness in the environmental policies sector was not at all displeasing to the political parties: in a word, one gets the clear impression that environmental commitment had the function of silencing the functionaries and offices assigned to studying ecological problems, as well as those minority sectors of science and society that had begun to speak out about the ecological crisis brought on by postwar development. Environmental rhetoric, in brief, assumed the function of communications maquillage directed at public opinion, while formal actions were still solidly centered around questions of work and production, to which ecological issues were irremediably subordinated.

The four essays comprising this first part (chapters 1–4) are dedicated to the uses of natural resources. The first, which deals with the development of renewable energies, shows how even in a region like Tuscany—which has been seeking for decades to reconvert its economy by investing in renewable energy, experiential tourism, and quality agriculture—it is very difficult to apply an economic model capable of freeing itself from fossil fuels. The second and third chapters tackle resource management in the context of the region's main urban area (Florence-Prato). For all that the essays are not exhaustive, they do represent an attempt to show how one of the chief interpretive keys of environmental transformations is that of conflict. In Tuscany, as in other Italian urban areas,[6] the construction and organization of the region are seen to be the complex outcome of cooperation and competition among the interests of industry, citizens, various economic players, the cities themselves, their minor satellite towns, and the rural areas at the margins of modernization processes. Chapter 4 looks at the coast, where economic development and environmental transformations have been driven by tourism and by certain industrial sites in areas suitably equipped with water resources and seaports. In full awareness of the paucity of historiographical research and reflection in this field, the essay sets itself the task of first focusing on a sector of investigation (industrial development of coastal areas and, more generally, of high landscape value sites) whose study could be very useful in the analysis and measurement of conflicts triggered by the presence of heritage sites in areas selected for the building of large industrial plants. Adorno and Neri Serneri have in fact written that the environmental movement in Italy has a pronounced humanistic imprint, and this has led to thinking of the environment not in terms of ecosystem but essentially as a landscape (i.e., a cultural representation of the natural world).[7] To understand this characteristic trait of Italian environmentalism—still deeply rooted today—one would need to go into depth on the interactions between Fordist industrialization, environmental imbalance, and growth processes based on tourism and agriculture, not to mention the reconversion projects following the deindustrialization that took place from the end of the 1980s through the 1990s.

CHAPTER 1

THE ECONOMIC AND ENVIRONMENTAL IMPACTS OF BIOMASS, SOLAR AND MINI-WIND POWER PLANTS

THE WORLD, ITALY, AND TUSCANY

In the initial years of the twenty-first century, the production and consumption of energy are still firmly characterized by a predominant use of fossil-based energy sources. According to assessments made by the World Resources Institute, in 2001 the global consumption of energy was distributed as: 79.5% fossil fuels, 10.4% solid biomass, 6.9% nuclear, 2.2% hydroelectric, and 0.7% other renewables. In Europe energy consumption was characterized by a higher use of fossil fuels (84.2%) and nuclear (10.5%), which relegates the percentage of renewables to a mere 4.7% (2.4% hydroelectric, 2.0% solid biomasses, 0.3% other renewables).[1] In 2004 only 0.51% of world energy was derived from solar or wind.[2]

In 2005 renewable sources (including nuclear) represented 10.2% of global energy consumption according to OPEC, and 7.68% according to the Energy Information Administration of the US Department of Energy.[3] *World Energy Outlook 2008* of the International Energy Agency envisages the future of energy still dominated by fossil fuels:

> In our Reference Scenario [policies and regulations adopted by mid-2008], world primary energy demand increases by an average of 1.6% per annum during the period between 2006 and 2030, increasing from 11,730 million tons of oil equivalent (toe) up to just over 17,010 million toe, with a growth of 45%. . . . In 2030, fossil fuels account for 80% of global primary energy mix,

> which is slightly lower than the current level. The oil continues to remain the predominant fuel, despite the demand for coal increases more than that of any other fuel in absolute terms. The percentage of world energy consumed in urban centers, estimated at 7,900 million toe in 2006, increased by two-thirds to almost three quarters in 2030.[4]

According to the International Energy Agency, however, renewable technologies are destined to become "the second most important source for the generation of electric power after coal":

> Excluding biomass, non-hydroelectric renewable sources such as wind, solar, geothermal and tidal grow overall more than any other source worldwide, at an average rate of 7.2% per annum for the period in question. Most of the increase occurs in the electricity sector. The percentage of non-hydroelectric renewables to total electricity production increases from 1% in 2006 to 4% in 2030. The production of hydroelectric power is growing, even though its share of electricity production fell by two percentage points to 14%. In OECD countries, the increase of renewable sources for generating electricity exceeds that of fossil fuels and nuclear power combined.[5]

Even the *International Energy Outlook 2008* of the Energy Information Administration confirms the future sustained growth of renewables (2.1% per year between 2005 and 2030) and, for 2030, predicts a scenario that provides a world energy consumption still very dependent on fossil fuels (Graph 1.1). Hydropower and other renewables would grow from 923 megawatts (MW) in 2005 to 1,373 in 2030, with an average annual growth of 1.6% (+48.75% overall). In Europe the installed capacity would rise from 212 to 293 MW (+1.3% per annum, +37.73%).[6]

In Italy between 1995 and 2005 the availability of energy sources rose from 171,716 million tons oil equivalent (toe) in 1995 to 197,776 in 2005.[7] The period in question was characterized by a progressive decrease in oil consumption and by a constant increase in natural gas consumption. In addition, energy policies have been characterized by an attempt to stimulate the development of renewable sources and to encourage energy saving measures. In 2005 the final consumption of energy amounted to 146,591 million toe: 69,219 oil (47.22%), 45,050 natural gas (30.73%), 25,866 electricity (17.65%), 4,629 solid fuels (3.16%), 1,827 renewables (1.25%). Of the 303,672 million kilowatt hours (kWh) of electricity produced, 83.33% were of thermoelectric origin, 14.13% hydroelectric, 1.75% geothermal, 0.77% wind, and 0.02% solar. The gross production of electric power from renewable sources was: 72.28% hydroelectric, 12.34% biomass and waste, 10.67% geothermal, 4.69% wind, and 0.02% solar.

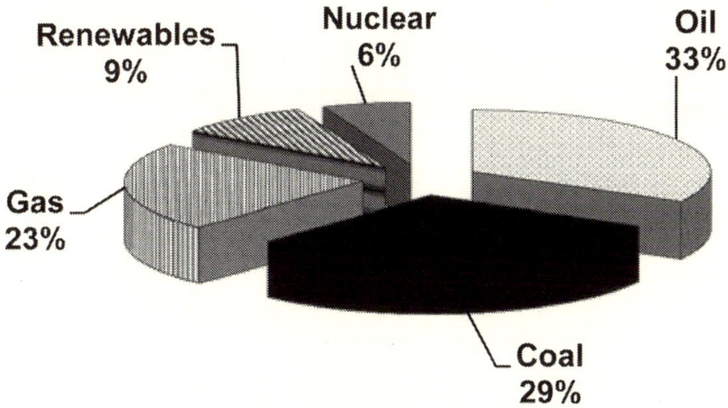

Graph 1.1. World total energy consumption by fuel (%); projections for 2030. *Source:* EIA, *International Energy Outlook 2008* (Washington, DC: EIA, 2008).

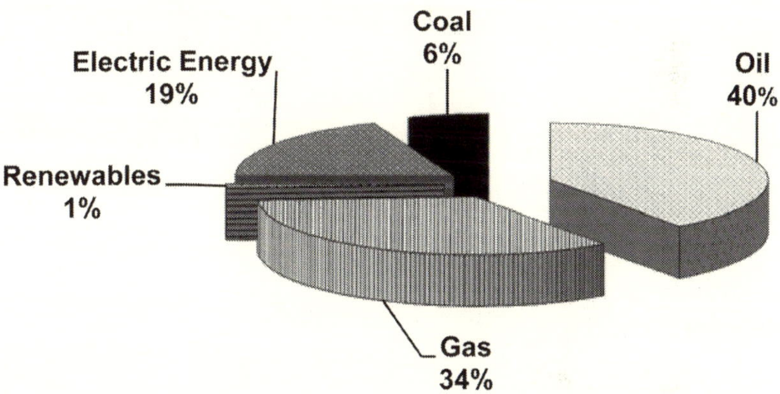

Graph 1.2. Total energy consumption by fuel (%) for Tuscany, 2004. *Source*: ARPAT, *Relazione sullo stato dell'ambiente in Toscana 2008* (Florence: Edifir, 2008).

After a significant reduction of consumption during the final years of the twentieth century, the demand for energy in Tuscany—led by industry and transportation sectors—began to increase again in 2001, with a mean annual rate more than double that of the 1990s. In 2004 Tuscany's final consumption amounted to 6.74% of the national, largely matching its composition, with a clear prevalence of fossil fuels, in particular oil and natural gas (Graph 1.2).[8]

Forecasts for the country in the *Libro bianco per la valorizzazione energetica delle fonti rinnovabili* (April 1999) gave as a target for the period 2008–2012 an increase in the use of energy from renewable sources up to about 20.3 million toe (Mtoe), as compared to 11.7 Mtoe recorded in 1997. In 2008–2012 electricity produced from renewables (16,744 Mtoe) was: 10,362 Mtoe hydro-

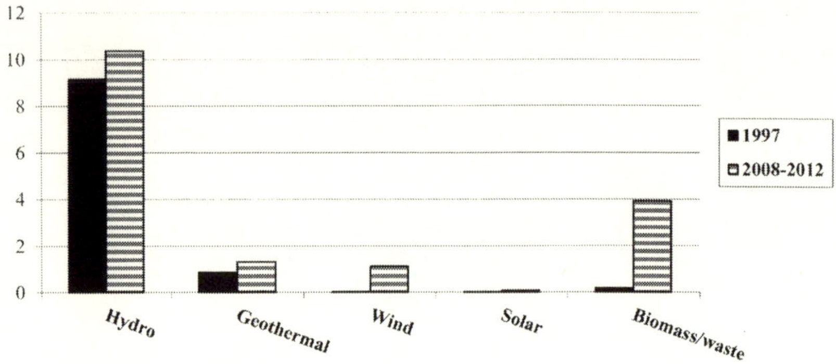

Graph 1.3. Electricity production (Mtoe) by renewables for 1997 and projections for 2008–2012. *Source*: ENEA, *Libro bianco per la valorizzazione energetica delle fonti rinnovabili* (Rome, 1999).

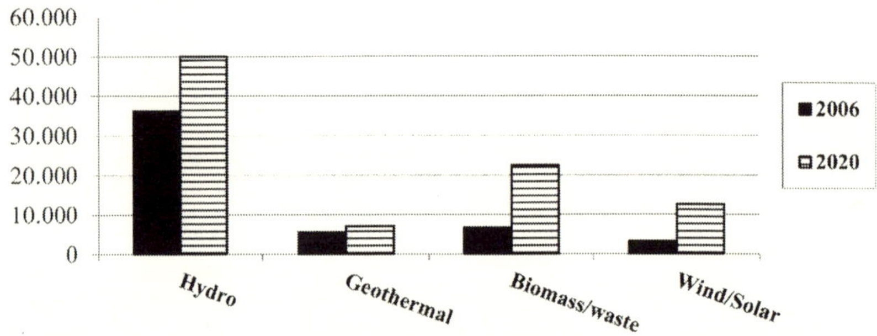

Graph 1.4. Energy demand in Italy (in GWh) for renewables, 2006 and projected for 2020. *Source*: Unione Petrolifera, *Previsioni di domanda energetica e petrolifera italiana 2007–2020* (Rome, 2007).

electric, 3,916 biomass and waste, 1,294 geothermal, 1,100 wind, and 0,073 solar (Graph 1.3).[9]

The Unione Petrolifera speculates that between 2006 and 2020, there will be a considerable decrease in the demand for oil as a proportion of the total energy demand (from 44.1% in 2006 to 35.2% in 2020) in favor of natural gas and renewables (41% and 10.8%, respectively, in 2020). The Unione Petrolifera foresaw for 2020 a production of 92,000 gigawatt hours (GWh) (Graph 1.4): 50,000 GWh hydroelectric (+38.35% compared to 2006), 22,500 biomass (including urban waste, +235.82%), 12,500 wind-solar (+288.80%), and 7,000 geothermal (+26.65%).[10]

WIND AND MINI WIND POWER

Between the mid-1990s and 2007, wind energy experienced significant development, gaining important sectors of the market in Germany, Denmark,

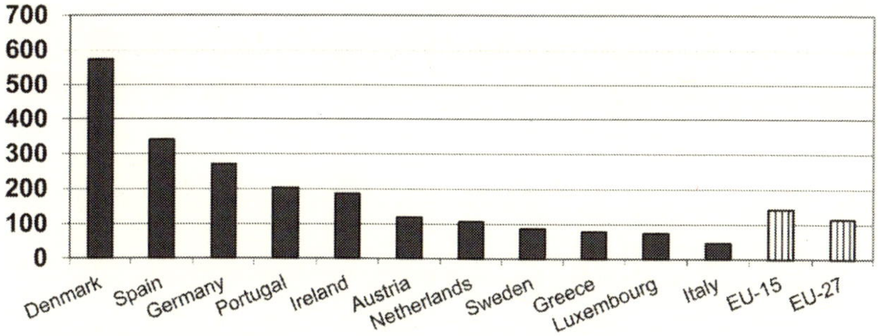

Graph 1.5. Wind power capacity (kW/1,000 inhabitants) for EU-27, 2007. *Source*: European Wind Energy Association, *Wind Energy in Europe* (Brussels, 2007).

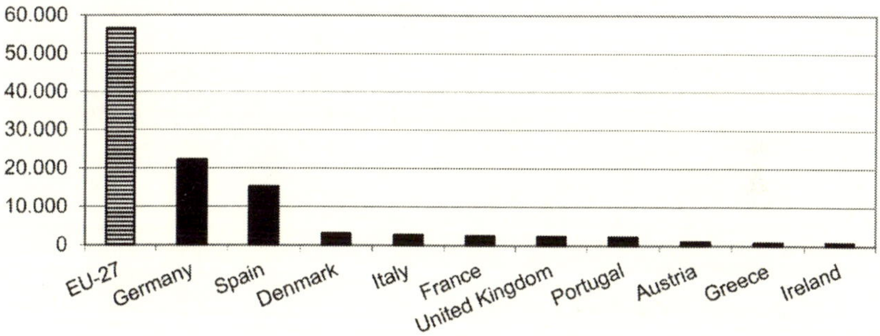

Graph 1.6. Wind power capacity (MW) for EU-27, 2007. *Source*: European Wind Energy Association, *Wind Energy in Europe* (Brussels, 2007).

Spain, Portugal, and Ireland.[11] These five countries (Graph 1.5) boasted the greatest wind capacity within the European Union, above the average of the EU, both of EU-15 and EU-27. Germany, Spain, and Denmark—the first-adopters of wind power—continued to attract the highest investments, followed by France and Portugal. Wind power contributes to over 21.22% of electricity production in Denmark, 11.76% in Spain, 9.26% in Portugal, 8.42% in Ireland, and 7% in Germany. At the end of 2007 the wind power capacity installed in EU-27 was 56,535 MW: Italy had occupied fourth place, but dropped to eleventh considering the relationship between installed power and number of residents. Italy appeared to be far behind on offshore wind farms, with a capacity of 0.08 MW, compared to 590.8 MW in the United Kingdom and 426.35 in Denmark, a fact of primary importance if we consider that the European Wind Energy Association speculates that in 2015, offshore wind farms will achieve a total power production almost equal to 31,000 MW (Graph 1.6).

The success of wind energy, despite fixed costs being generally greater than that of fossil fuels (the cost per kWh varies from €1,033 for the European inland to €1,550 offshore),[12] has been established by the low variable costs due to low operating costs and to the fact that the wind spares the producers the high cost of oil. According to the task force for renewable energies of the G8, wind is also the energy source with fewer externalities (the consequences of an industrial activity which affects other parties without this being reflected in market prices; for example, manufacturing activities that cause air pollution impose health and cleanup costs on all of society).[13]

It's precisely the above-mentioned factors that have encouraged the spread of large wind farms, such as the Middelgrunder farms (20 offshore turbines) and Syltholm in Denmark (35 inshore turbines of 750 kW). Beginning in 1995 wind farms have also spread in Italy, where in 2000 there were fifty-eight active plants generating installed power of about 430 MW.[14] In 2007, 157 Italian municipalities with wind plants generated a total of 2,819 MW. The main plants are situated in the regions of Toia (167.9 MW), Agrigento (104.55 MW), Bisaccia (101.88 MW), Sant'Agata di Puglia (97.2 MW), and Ulassai (84 MW).[15]

Despite wind power, along with biomass, being regarded as the principal alternative to fossil fuels and an essential instrument for meeting the European goals for 2020 (20% of total energy obtained from renewables),[16] wind farms have become one of the chief targets of Not in My Backyard (NIMBY) protests. The greatest worries concerning wind farms regard the occupation of land, noise, electromagnetic waves, visual impact on the landscape, and interferences with birds (the best known case being the Altamont Pass in California) and, in the case of offshore plants, with marine fauna.[17]

Studies carried out to date, however, appear to partly assuage the concerns expressed by a section of the environmental movement. In Denmark, for example, studies have been made on onshore installations of Tunø Knob and offshore Vindeby: in the first case "no difference has been noted in the behavior of birds while the rotors were in action," and in the second "the result has been the realization of an increase in fish catches, after the building of the plant, due to the fact that the foundations of the windmills (the gravity ones) are seen by the fauna as an artificial reef."[18] In Italy the case study on the Serra Lunga plant at Cinisi (Palermo) assessed the consequences of the plant:

> Despite the location of the plant being in the "SIC/ZPS Montagna Longa e Pizzo Montanello," in the light of the data . . . regarding the effects of the ongoing installation works on the local species of flora, vegetation and habitat, or altogether environmental effects, we can say that the impact of the emplacement of the turbines will be more evident in quantitative than in qualitative terms and only in the short term, since the species and habitats of higher value are connected with floral communities for centuries adapted to heavy human

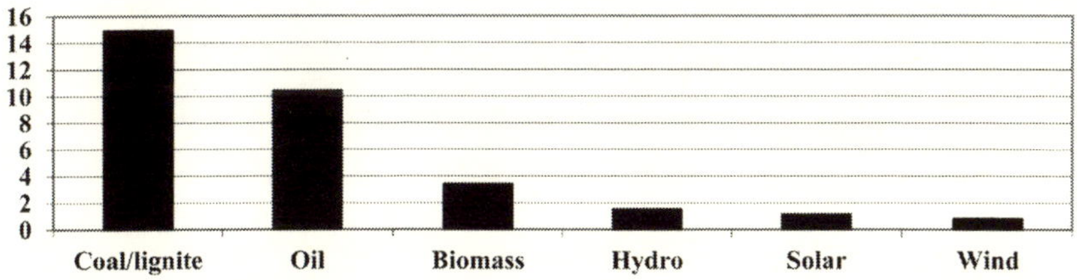

Graph 1.7. Estimates of the externalities (Euro cents/kWh) of electricity production. *Source*: Pietro Menna, *L'energia pulita. Sole, acqua, vento . . . le fonti di energia rinnovabili* (Bologna: Il Mulino, 2003), 85.

> interference. For what concerns flora, vegetation and fauna, the overall impact of the plant appears negligible.[19]

A study carried out by the Centro Ornitologico Toscano for the Regione Toscana has found that the most significant environmental impacts caused by wind farms concern large birds (the hazard of collisions with the turbines has been described as "a limiting factor for the conservation of bird populations") which, along with the "disturbance caused by ordinary and extraordinary maintenance," tends to induce birds of prey and sparrows to abandon the sites where the wind plants are erected. For this reason the study recommends that turbines not be placed in areas where "particularly rare and delicate species exist," or near wetlands, reservoirs and "along the ridges of hills (watershed and areas immediately adjacent to it) and the same for mountains, in particular at a high altitude," since winds "are stronger and capable of diverting the flight of the birds."[20]

In Italy concerns about those impacts have changed opinions about wind power from an indispensable resource for reducing climate-changing emissions to an environmental problem. *Piano di indirizzo energetico regionale* declares that in Tuscany "wind power has suffered from the climate of prejudice and hostility which has spread against large wind generator plants chiefly for worries about excessive impact on the landscape."[21]

The broad dissent toward the installation of wind turbines is not a simple problem if we consider that the Regione Toscana regarded wind power a strategic resource.[22] Over the last ten years, with the consent of environmental associations, the use of what is called *mini wind* (small wind turbines, not taller than 30 m) has begun to take off. These plants are fairly well distributed in Great Britain, where they have been supported by the British Wind Energy Association since the early 1990s.[23] In Italy the first experimental mini-wind plant came online in Trento in 2007 (three turbines with a power of 1 kW, 11 kW, and 20 kW).[24] Compared to larger wind farms, the mini wind is more

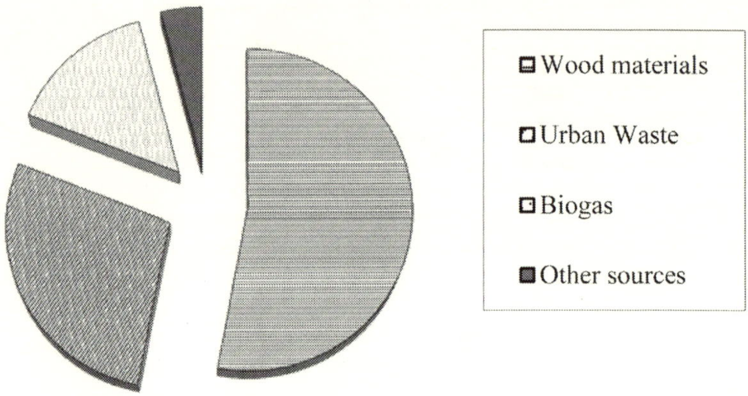

Graph 1.8. Percentage of biomass energy production by different sources for EU-25, 2005. *Source*: European Biomass Industry Association, *Annual Report* (Brussels, 2007).

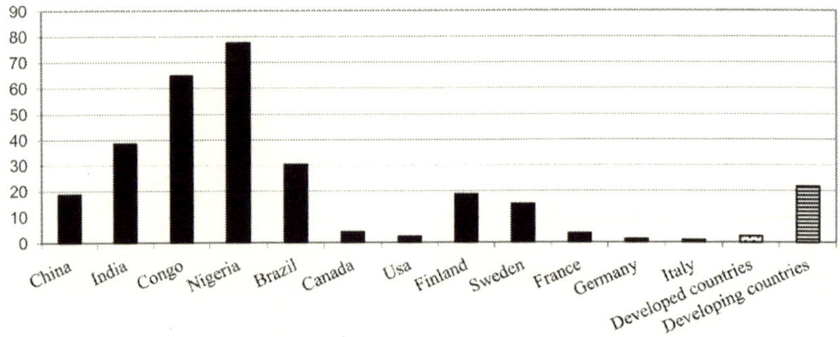

Graph 1.9. The percentage of biomass in total energy consumption, 2001. Source: World Resources Institute, *World Resources 2005: The Wealth for the Poor.* Managing Ecosystems to Fight Poverty (Washington, DC: World Resources Institute, 2005), 200–202.

easily integrated into the landscape, and in the future may make it possible to create a self-productive network, a "network of agro-energetic businesses, small business districts both productive and sustainable, which in part directly utilize the energy produced and partly exchange it in a network, with investments giving a return in a matter of few years."[25]

BIOMASS

As we have seen, biomass is the most used renewable source for the production of energy. This is because in many developing countries, firewood is still the cheapest fuel and the easiest to obtain (Graph 1.8).

In developed countries, the waste produced by the wood industry constitutes over 50% of the biomass employed in energy production: in the United

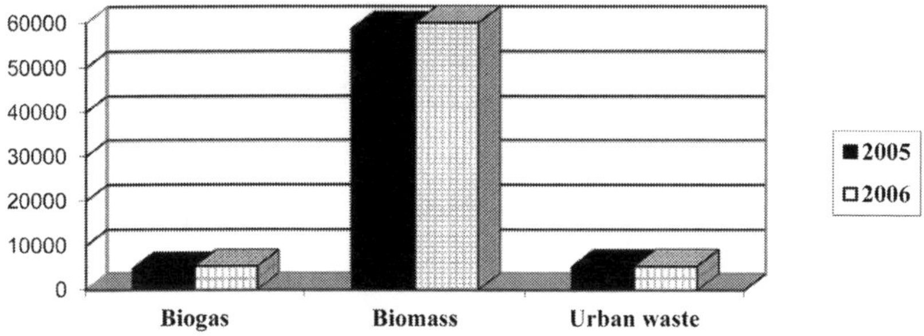

Graph 1.10. Energy production by biomass (kToe) for EU-25, 2005, 2006. *Source*: European Biomass Industry Association, *Annual Report* (Brussels, 2007).

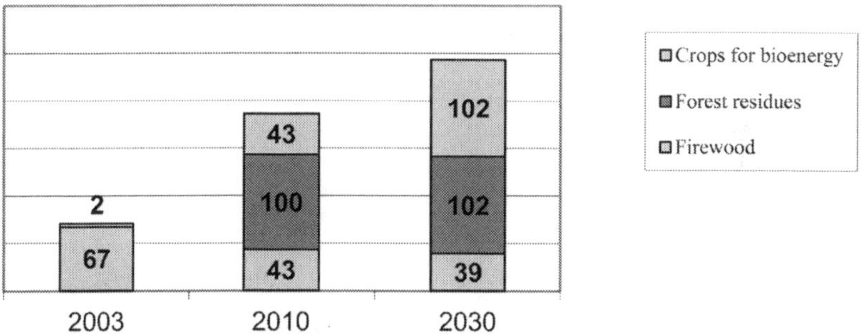

Graph 1.11. Energy production by biomass (Mtoe) for EU, 2003 and projections for 2010 and 2030. *Source*: Commission of the European Communities, *Comunication from the Commission: Biomass Action Plan*, December 7, 2005.

States 10,000 MW are derived from the combustion of forestry waste, and St. Paul, Minnesota, has a district heating plant which utilizes 250,000 tons of wood industry waste per annum in place of coal, with a drop in carbon dioxide emissions of 76,000 tons per year.[26] In Europe, Finland and Sweden are the leading countries employing wood industry waste, whereas France, Germany, and Austria are the leading countries in the production of energy by the incineration of urban waste (Graphs 1.9 and 1.10).[27]

The combustion of these biomasses is carbon neutral because the carbon dioxide producted during combustion is equivalent to that absorbed during the growth of the same biomass.[28] The European Union foresaw a substantial increase in the use of this energy source in 2010, according to a scenario that, at least hypothetically, excluded the negative effects on food production on the loss of biodiversity, and on the use of soils and water resources (Graph 1.11).[29] In addition, the EU maintains that an increased use of biomass could ensure an increase in employment in rural areas (250,000–300,000 new jobs) beyond contributing to a reduction in energy imports (between 42% and 48% less) and greenhouse gases (a decrease of 209 million tons equivalent of carbon dioxide/year). According to the estimates of the European Commission, the increased power of biomass energy systems would cost €9 billion/year, 6 billion of which would be absorbed by biofuels and 3 billion by solid biomass.

Despite the optimism expressed by the EC (shared by some important research centers),[30] the use of edible vegetables for the production of biofuels could upset food markets, as has happened for corn, whose jump in price—though certainly also induced by financial speculation—has caused an increase in the price of many products (from tortillas in Mexico to pasta in Italy). At present the demand for corn for human food and animal feed grows annually at the rate of 1%, whereas the demand for bio-ethanol grows by 20%. Some recent studies have shown how the conversion of forests and grazing areas into fields for the production of biofuels may actually generate from 17 to 420 times the annual emissions of carbon dioxide, although it was believed this could be mitigated by their replacing fossil fuels.[31]

Furthermore, the use of biomass, although presenting decidedly positive externalities compared to fossil fuels, has a higher impact than solar and wind energies, because nitrogen oxide emissions produced during cultivation, harvest, transportation, and combustion are less than those of fossil fuels. A study by the Italian National Agency for New Technologies, Energy and Sustainable Economic Development (ENEA) recommends the realization of biomass power plants "only in contexts of high production of forestry refuse, at distances of supply around 100 km from the plant." Wherever such conditions do not exist, ENEA suggests that wood residue should be employed for the "production of local heat and steam for local use." In addition, "the forced production of biomass on a large scale (short rotation forestry) creates chiefly environmental problems which at the moment are still to be assessed."[32] In 2004 in Italy biomass provided around 32% of all energy produced with renewable sources (Graph 1.12).

The report *Comuni rinnovabili 2008* by Legambiente studied 306 municipalities which have a biomass plant in their region, for 770 MW total power. In addition, 267 municipalities possess tele-heating plants (for the production of heating and hot water) that serve 460,000 users, generating 5,000 GWh/year and heating more than 1,217 m^3 of water.[33]

According to the Ministry of the Environment, the availability of residual biomass amounts to around 66 million tons of dry matter, equal to 27 Mtoe; this means that "with an adequate program of reforestation and man-

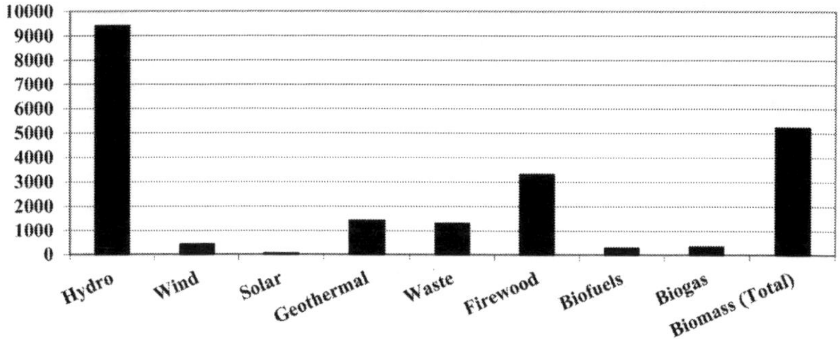

Graph 1.12. Italy. Renewable energy production (equivalent fossil energy replaced, ktoe) for Italy, 2004. *Source*: ENEA, *Rapporto energia e ambiente 2005: I dati* (Rome: ENEA, 2005).

agement of forests we may achieve an availability of 2 Mtoe/year." In addition, "we might obtain new conventional woodlands by the reforestation of part of the over 2 million hectares not destined for agriculture because they're not very productive, and deciduous forests may be planted, along with grasses, for energy production, reclaiming part of the 250 thousand hectares now left fallow."[34]

Just as with wind energy, biomass energy production is opposed by NIMBY protesters: of 172 recent protests, 50 (29%) have opposed waste-to-energy plants and 5 (2.9%) biomass plants. The diffidence toward this energy source chiefly concerns their effects on health, because citizens do not always understand the mathematical models that are used for assessing allowable concentrations of pollutants and believe that administrative controls are not adequate for preventing "the introduction of matter other than wood chips."[35] In Tuscany in 2003 biomass contributed 0.13% to the production of electricity, compared to 0.02% for wind energy and 0.01% for solar photovoltaic. Biomass, whose annual availability is estimated at 2.5 million tons, is employed for the heating of greenhouses, for domestic heating, and for recycling industrial waste. The *Piano di indirizzo energetico regionale* assumed an installed capacity of 180 MW for a production capacity of 1,025 GWh, an objective based on the fact that Tuscany is the Italian region with the greatest extent of forests and annually produces 295,912 m^3 of wood for energy use (15% of the nation's total) employed in low-yield systems (i.e., chiefly in stoves and fireplaces).[36]

SOLAR PHOTOVOLTAIC SYSTEMS

Together with wind, solar energy is the source with the fewest externalities (see Graph 1.7). During their life cycle (twenty-five years on average) solar panels do not present risks to the natural environment or to human health.

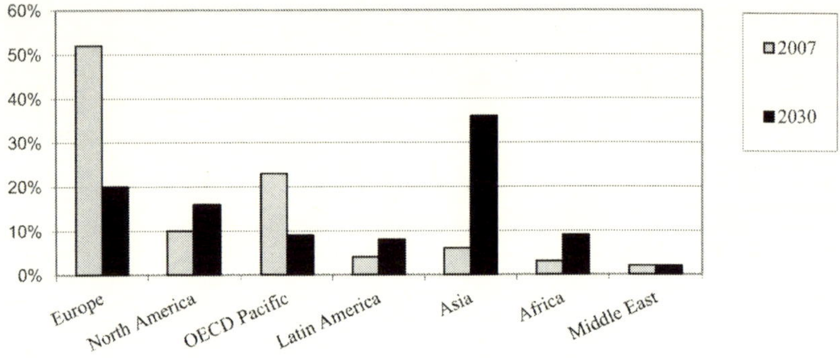

Graph 1.13. The proportion of installed photovoltaic systems, 2007, and estimates for 2030. *Source*: EPIA and Greenpeace, *Solar Generation V-2008: Solar Electricity for over One Billion People and Two Million Jobs by 2020* (Brussels: EPIA; Amsterdam: Greenpeace, 2008).

Their only negative aspect may be the occupation of land and the possible interference with urban landscapes (solar panels are installed on roofs or walls). The most significant environmental impacts occur during the production of the panels, which employ toxic substances such as silane, phosphine, diborane, hydrogen selenide, and cadmium. Panels also represent a special waste product, whose disposal necessitates the separation and recovery of their toxic metals.[37]

Until the end of the 1990s solar energy was chiefly utilized for the production of hot water through thermo-accumulators. In China in 2008, 124 million m^2 of solar accumulators were in operation (with an output comparable to fifty-four coal plants) providing hot water even in villages without electricity infrastructure.[38] In Europe the primary users of solar panels were Austria (15% of Austrian homes were provided with solar panels for heating water) and Germany. By 2020 it is estimated that about 1,080 million m^2 of solar thermal collectors will be in operation: 500 m^2 in Europe, 300 in China, 200 in the United States, and 80 in Japan.[39]

Sunlight is also used to produce electricity by heating fluids to high temperatures (400°C). This concentrated solar power technology has been developed mostly in California, where between the 1980s and 1990s concentrated solar power plants have become operational with a capacity of more than 350 MW. In 2007 a 64-MW concentrated solar power plant was built in Nevada, whereas in Spain and Florida two more such plants have been studied and designed.[40] However, the real "boom" has been in photovoltaics, or the conversion of sunlight directly into electricity through so-called solar cells.

The overall capacity of photovoltaic systems has increased from 1,200 MW in 2000 to 9,200 MW in 2007, an increase of 766%.[41] In 2007, 85% of photo-

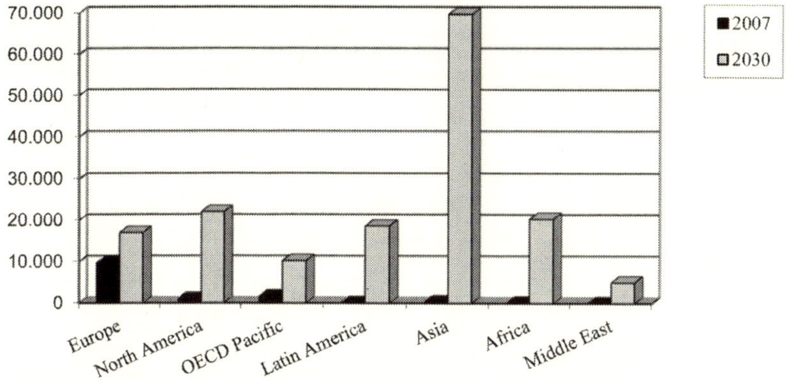

Graph 1.14. Market value (in millions of euros) of photovoltaic systems, moderate scenarios for 2007 and estimates for 2030. *Source*: EPIA and Greenpeace, *Solar Generation V-2008: Solar Electricity for over One Billion People and Two Million Jobs by 2020* (Brussels: EPIA; Amsterdam: Greenpeace, 2008).

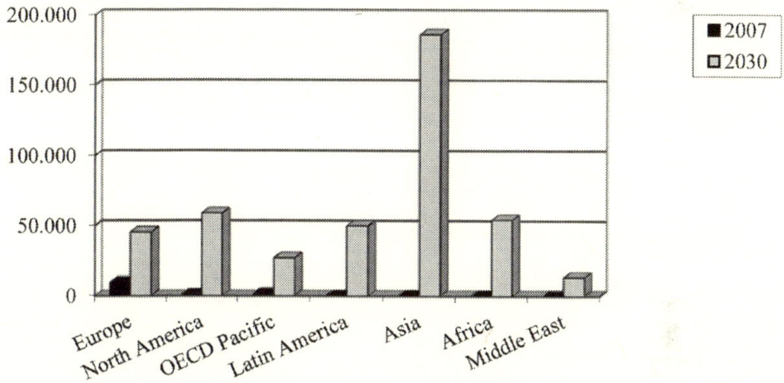

Graph 1.15. Market value (in millions of euros) of photovoltaic systems, advanced scenarios for 2007 and estimates for 2030. *Source*: EPIA and Greenpeace, *Solar Generation V-2008: Solar Electricity for over One Billion People and Two Million Jobs by 2020* (Brussels: EPIA; Amsterdam: Greenpeace, 2008).

voltaic systems were installed in Europe, North America, and the Pacific, of which 73% was concentrated in Europe, where the photovoltaic sector was driven primarily by investments in Germany and Spain (Graphs 1.13–1.15).[42]

As can be seen from Graphs 1.14 and 1.15, the European Photovoltaic Industry Association (EPIA) suggests two growth scenarios under which, in 2030, the photovoltaic market will be dominated by Asia, Africa, and Central and South America, which together hold an estimated share of around 64%. According to the EPIA, the developing countries will be able to become the reference market for photovoltaic technology. In particular, China will take the lead, with investments almost double those of Europe.

The EPIA estimates that in 2030 installed capacity will reach 912 GW with a production of 1,291 terawatt hours (TWh), equivalent to 6.73% of the total worldwide consumption of electricity (13.79% in the advanced scenario). The success of photovoltaic systems also would create a significant number of new jobs (between 3 and 9 million) and contribute significantly to the abatement of greenhouse gas emissions (Graph 1.16). These scenarios assumed a continued reduction in the cost of solar power, that in 2007 ranged between €0.44 and €0.22 per kWh. In 2030 producing a kWh should cost between 60% and 70% less than in 2007 (Graph 1.17).

According to Lester Brown, for those who live in areas not connected to a power supply, it "is often cheaper to install photovoltaic panels on the roof rather than building a centralized system and a network to reach potential consumers."

> For the inhabitants of the villages in the Andes, who have always used tallow candles for lighting, the monthly cost to install a solar panel system with a 30-month payment plan is less than the monthly cost of the candles. Indian villagers who are not reached by the electricity grid and depend on kerosene lamps, face similar charges. Installing a home solar system in India costs about $400. These systems feed two, three or four small appliances or lighting points and are widely used in homes and shops instead of kerosene lamps, which are polluting and increasingly expensive. In one year, a lamp burns about 75 liters of kerosene, which at 80 cents per liter means $60 per lamp. A system of solar cells, which replaces two lamps, pays for itself within four years.[43]

The city of Rizhao in Shandong Province, China (about 3 million residents), is one of the first examples of a "solar city." Solar panels provide hot water to 99% of the families who live in the central districts and 30% of those living in the suburbs. In addition, six thousand families use solar energy in the kitchen and sixty thousand greenhouses are heated by means of photovoltaic systems. Overall, Rizhao has installed 500,000 m^2 of solar collectors.[44]

Following the example of Barcelona, over seventy Spanish cities have adopted building rules requiring compulsory installation of photovoltaic systems in completely renovated and new buildings. The city of Sacramento, California, have distributed monetary incentives to residents who install photovoltaic panels, with the goal of achieving a rate of 23% of total electricity consumption coming from renewable sources. In Daegu, South Korea, the Center for Solar City Daegu has promoted, in collaboration with the local

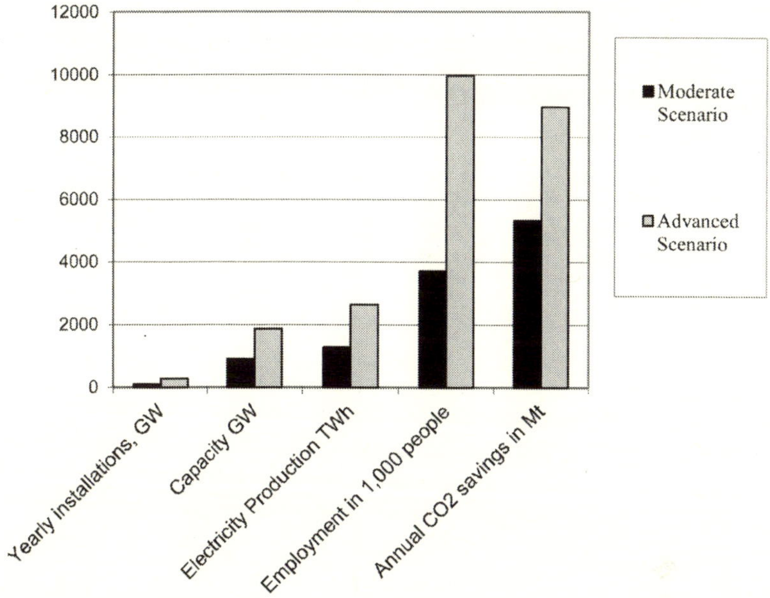

Graph 1.16. Estimated global photovoltaic generation (Mtoe) scenarios for 2030. *Source*: EPIA and Greenpeace, *Solar Generation V-2008: Solar Electricity for over One Billion People and Two Million Jobs by 2020* (Brussels: EPIA; Amsterdam: Greenpeace, 2008).

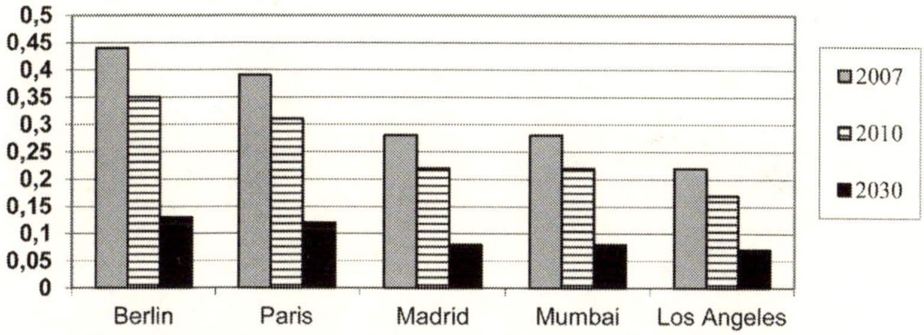

Graph 1.17. Estimated photovoltaic generation costs (in euros) per kWh for 2007 and projected for 2010 and 2030. *Source*: EPIA and Greenpeace, *Solar Generation V-2008: Solar Electricity for over One Billion People and Two Million Jobs by 2020* (Brussels: EPIA; Amsterdam: Greenpeace, 2008).

administration, the installation of photovoltaic systems in school buildings, on campuses, and in water treatment plants. The cost of installing solar panels on house roofs was subsidized at 80% by local government, in addition to national subsidies. In Mexico City the World Bank and some nonprofit organizations have launched a project to install fifty thousand solar central heating systems. Finally, in Cape Town, South Africa, the Clean Development Mecha-

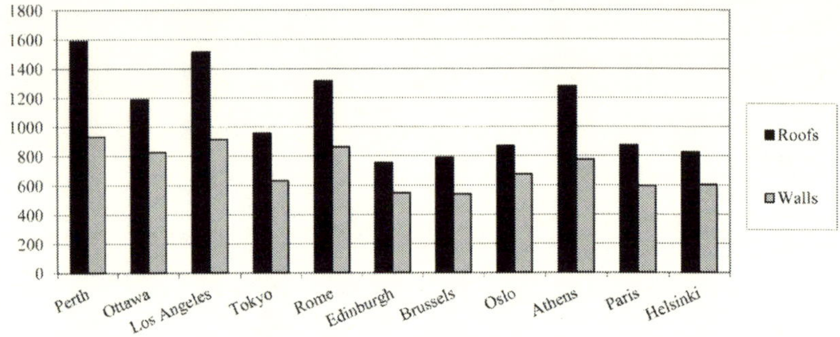

Graph 1.18. Production of electricity by photovoltaic systems(kWh/kW year). *Source*: International Energy Agency, Photovoltaic Power Systems Programme, *Compared Assessment of Selected Environmental Indicators of Photovoltaic Electricity in OECD Cities*, IEA PVPS TASK 10, ACTIVITY 4.4., REPORT IEA-PVPS T 10–01:2006, May 2006.

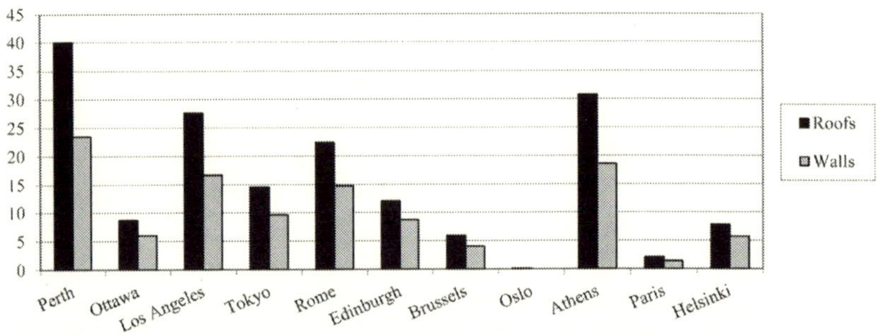

Graph 1.19. Carbon dioxide savings for photovoltaic systems (tons/kW). *Source*: International Energy Agency, Photovoltaic Power Systems Programme, *Compared Assessment of Selected Environmental Indicators of Photovoltaic Electricity in OECD Cities*, IEA PVPS TASK 10, ACTIVITY 4.4., REPORT IEA-PVPS T 10–01:2006, May 2006.

nism of the Kyoto Protocol has promoted a pilot project for the distribution of solar water heating systems to residents.[45]

A study sponsored by the International Energy Agency shows that the cost of installing a photovoltaic system can be paid for on average between 1.6 (Perth) and 3.3 years (Edinburgh) for those installed on roofs and between 2.7 (Perth) and 4.7 years (Brussels) for those fitted on walls.[46] The annual energy production of rooftop systems can vary between 1,587 kWh/kW year for Perth and 754 for Edinburgh, with a decrease in carbon dioxide emissions of between 40 tons/kW (Perth) and 0.1 tons/kW (Oslo). For those systems installed on walls, the annual energy production can fluctuate between 932 kWh/kW in Perth and 539 in Brussels with a decrease in carbon dioxide emissions between 23.5 tons/kW (Perth) and 0 tons/kW (Oslo) (Graphs 1.18 and 1.19).

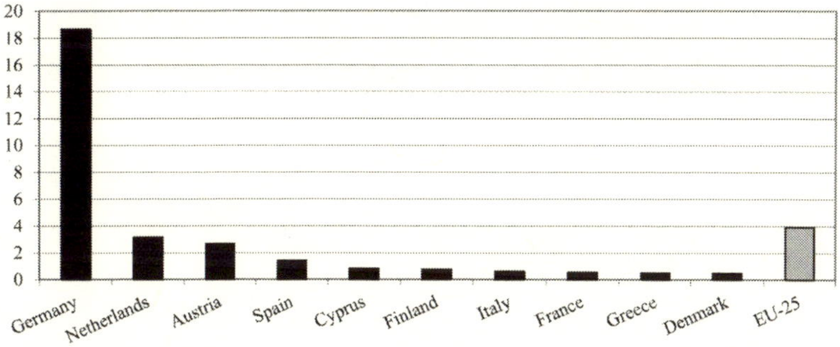

Graph 1.20. Solar photovoltaics in some European countries (Watt peak/resident). *Source:* EurObserver, *Photovoltaic Energy Barometer 2006* (Paris, 2006).

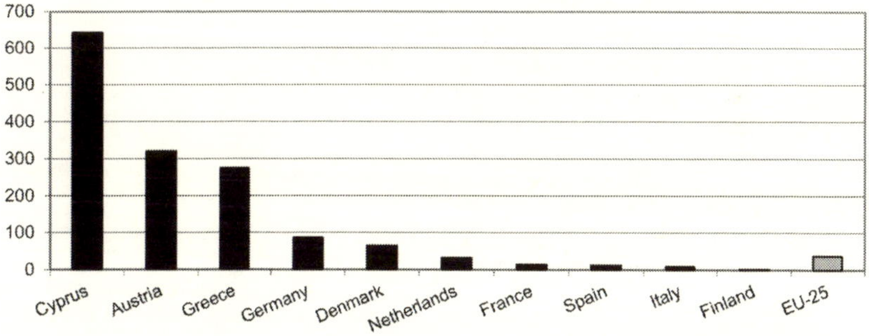

Graph 1.21. Solar photovoltaics in some European countries (m²/1,000 residents). *Source:* EurObserver, *Photovoltaic Energy Barometer 2006* (Paris, 2006).

Despite the great potential of Italy's location—the global horizontal irradiation is 1,552 kWh/m² in Rome and 1,251 kWh/m² in Milan (compared to 1,045 in Amsterdam and 972 in Cologne)—its delay in adopting solar power, compared to the more developed countries of the European Union, is even more stark.[47]

In Italy in 2004, solar thermal plants (460,000 m² of total area) contributed 0.60% to the generation of heat, while the share of electricity produced by solar photovoltaic was equal to 0.06% of the total (the installed capacity was 30,815 kWh).[48]

Solar systems were widespread mainly in Lombardy and Trentino–Alto Adige; the total area of solar thermal was a quarter of the European average, while the installed photovoltaic capacity was 6.5 times lower than the European average (Graphs 1.20 and 1.21).[49] Solar water heaters were present in 390 municipalities and photovoltaic panels in 2,799 (of which 1,339 had a population of less than five thousand). There were eighteen municipalities with installed

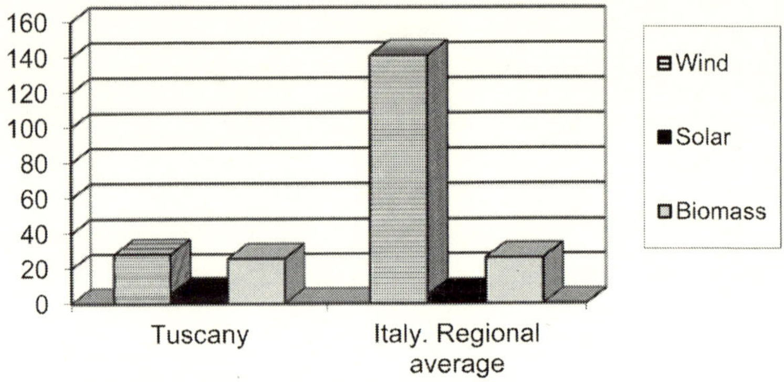

Graph 1.22. Renewable energies in Tuscany and Italy (MW), 2008. *Source*: Regione Toscana, *Piano di indirizzo energetico regionale (PIER): Proposta della Giunta Regionale* (Firenze, 2008).

capacity of more than 100 kW per one thousand residents.[50] In 2004 Tuscany had installed 22,500 m^2 of solar water heaters, while photovoltaic plants generated 4 MW (6.1 MW in 2008, according to Legambiente). The Regione Toscana expected to achieve a photovoltaic output of 45 MW, for a production capacity of 99 GWh by 2012, which corresponds to about 2% of the total production of electricity by renewables. Regione Toscana envisioned a system with a share of about 63% based mainly on geothermal and biomass energy.[51]

CHAPTER 2

THE ENVIRONMENTAL IMPACT OF URBANIZATION AND INDUSTRIALIZATION IN THE GREATER FLORENCE AREA

1945–2001

The character of the Florence-Prato conurbation[1] has been prompted by two series of reasons. The first concerns its environmental character: the area consists of hills and plains and is characterized by an important water system centered upon the Arno River (which crosses its entire territory) and upon seasonal but important tributaries. Thanks to its peculiarities, during the second half of the twentieth century this portion of territory has preserved, notwithstanding a high rate of population increase, a significant degree of biodiversity which spans from deciduous woodlands to wetlands.

The second concerns the nature of the development which occurred in the Florence-Prato plain, a change from a district of fields, villages, and small to medium-size towns to a conurbation with a strong industrial presence placed within a highly urbanized rural context. The area is highly suitable as the subject of a study examining the processes by which urbanization and industrialization have profoundly altered the use of resources, thus giving rise to an environmental crisis.

ECONOMIC DEVELOPMENT AND THE ENVIRONMENTAL CRISIS, 1946–1971

THE URBANIZATION PROCESS

The principal driving force of the environmental change is the tumultuous process of urbanization caused by massive immigration, both from the surrounding country and from the southern regions, to municipalities characterized by significant incipient industrial activities.

Under the pressure of an unprecedented migration, the reconstruction of Florence and other urban centers in the area was guided by an agreement which brought together the interests of builders, professionals in the construction industry, landowners, private speculators, and investment banks. Since 1946 the construction industry has been placed at the center of the local economics. The principal political and economic actors regarded urban development in the plain west of Florence—in the direction of the municipalities of Sesto Fiorentino and Prato—as a primary opportunity for growth for the entire economy of Florence. Urban growth was characterized by overstated forecasts concerning development expansion, by the lack of public services, by the conversion of environmentally privileged areas to high-quality private building development rather than to public greenspace, and by a system of transportation infrastructure favoring roads (planned without a clear notion of functional priorities) to the detriment of public forms of transportation. The future of Florence's urban structure was fatally established beginning in 1958, when a plan was approved allowing intensive urban development in the plain (with a density from 7 m^3 that could be built upon 1 m^2 up to 20 m^3/m^2). The plan allowed development in the hills while at the same time leaving the historical center without adequate conservation measures.[2]

The municipal administration tried to mend the disarray by passing a measure known as the Detti Plan of 1962—an urban plan promoted by Edoardo Detti, a university professor of architecture and councillor for urban development—which established as a priority the conservation on the historical center, reduced the density of buildings, allowed more space for public facilities and greenspaces, and redirected urban expansion in a northwesterly direction. Following the crisis which in 1964 led to the dissolution of the center-left town council, the Detti Plan was shelved without any opposition from the public: the numerous amendments to the plan, passed during the 1960s and 1970s, had favored, despite a modest rate of population growth, a very high rate of building expansion which almost doubled the extent of the city. A lack of public greenspace aggravated the problem: in 1972 population density allowed 3.62 m^2 of space per resident, a very small space which was also unevenly distributed. This situation was mirrored by the other municipalities of the province with a strong industrial vocation: Prato, Campi Bisenzio, Scandicci, Sesto Fiorentino, Calenzano, Lastra a Signa, and Signa. These municipalities expanded geometrically, thus generating a single conurbation where industrial settlements were mixed up with residential areas, the latter mostly inhabited by workers employed by the manufacturing industries.

During the mid-1960s the only commune (a small district) provided with a development plan was Impruneta. Furthermore, although since 1951 authorization to elaborate a comprehensive joint development plan of Florence (PIF)[3] had been requested for all municipalities, any attempt to begin such planning was ineffective because of continual quarrelling among the local governments of the hinterland—who were only interested in privileging new industrial plants and workers' housing on their own territory—and the Florence Town Council, which instead regarded the development of tertiary activities and residential housing as priorities. There was a political reason behind the municipalities' disagreement: from 1951 to 1957 Florence had been the only one administered by center and center-left councils led by Christian Democrat Party (Democrazia Cristiana).[4]

Notwithstanding hydrogeological and forestry obligations, urban development caused major negative consequences for the entire Arno basin (seriously affected, as it was, by the spread of industrial and residential housing)[5] and for the surrounding hills. The most blatant cases involved the Monte Morello area, which was deeply scarred by quarrying,[6] and an alluvial plain measuring 270 hectares (an area called *i Renai*, a sand quarry near a river) lying between the Arno and Bisenzio Rivers northwest of Florence. Until the 1950s this area had been made highly fertile by the silt deposited by the Bisenzio's regular floods and was intensely cultivated. In the course of the 1960s and 1970s, 210 hectares, rich in gravel of high quality under its soil, was subject to unchecked quarrying, causing severe environmental damage. Halted in 1978 by order of the mayor of Signa, quarrying had caused vast depressions, which turned into numerous ponds (today these have turned into valuable wetlands populated by herons and other wading birds).

THE INDUSTRIALIZATION PROCESS

The Florence-Prato area began to acquire the aspect of an important industrial district—with settlement characteristics not unlike those of the industrial areas of northern Italy—at the end of the nineteenth century. In 1892 an inquiry carried out by the Ministry of Agriculture, Industry, and Commerce recorded 557 factories in the Florentine district, employing 13,194 workers.[7] Industrial plants were located chiefly in the municipal territories of Florence (which hosted 49.55% of all factories and 72.92% of the industrial workforce), Fiesole, and Sesto Fiorentino.

About twenty years later, the industrial census of 1911 listed 3,839 factories employing 36,110 workers, primarily processing agricultural products (1,715 factories), textiles (722), and metals (692). In 1927 the manufacturing sector boasted 8,112 factories and 44,758 employees. Florentine manufacturers op-

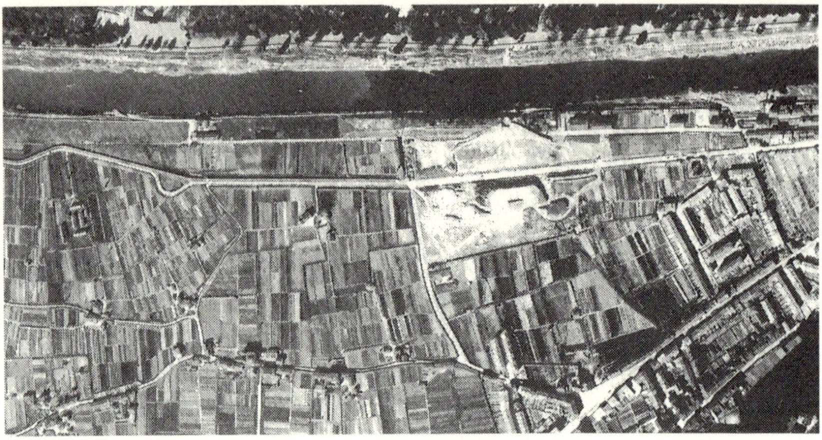

Figure 2.1. The Isolotto area—in front of the historical park of the Cascine, created by Cosimo I de' Medici—before urban development, 1935. *Credit*: Regione Toscana.

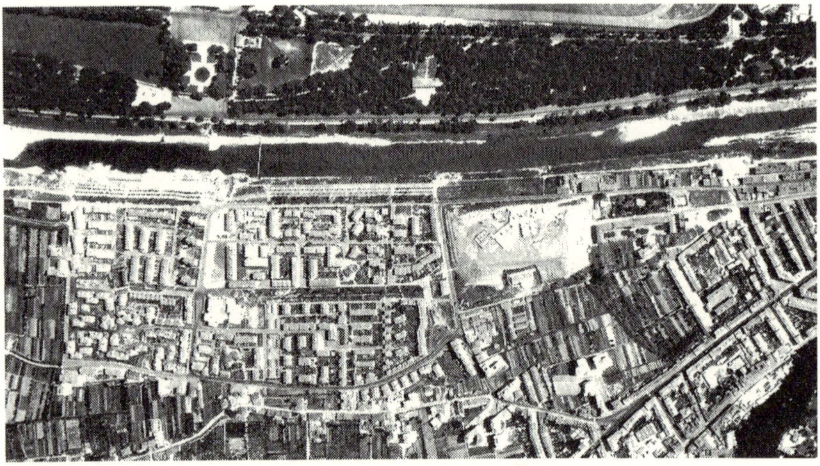

Figure 2.2. The Isolotto area after urban development, 1961. *Credit*: Regione Toscana.

erated chiefly in the clothing industry (3,016), mechanics (1,427), food (665), nonmetallic ores (438), textiles (239), and leather (175).

The quoted data show how by the first decade of the 1900s industrialization in the area had acquired a well-defined shape characterized by a manufacturing base and small plants (in 1927 plants averaged 6 employees). In the Prato area in 1888 the German conglomerate Kossler and Mayer set up a so-called *fabbricone* (big factory), a vast textiles plant which, with its one thousand workers, gave a decisive boost to industry already existing in the area. In 1927 the textiles sector employed 12,500 workers, 1,500 of which were in the fabbricone and over 1,000 at Forti (the area's largest firm based on local capital).

After the Second World War the entire area was subject to an intense period of industrialization, causing extraordinary growth throughout the municipalities of the Florentine belt[8] (meanwhile the municipal territory of Florence itself shifted to the service sector) and leading to the consolidation of small enterprises (the mean rate was 6.9 employees per local unit in 1951, 8.3 in 1961, and 7.0 in 1971).

In the Prato district war damages suffered by the major plants led to the fragmentation of industrial activity: damaged looms were bought up by former workers, who repaired them to start their own small businesses. Chronologically this process may be divided into two phases. Between 1951 and 1961 new plants tended to sprout up between Florence and Prato, and particularly in the municipal areas of Sesto Fiorentino and Calenzano. During this decade the number of employees rose from 58,815 to 90,235, of which 73,678 (81.65%) were employed in manufacturing plants. The number of workers in the Prato textile industry rose from 21,160 to 35,735.

Between 1961 and 1971 the lack of a plan for urban development led industrial expansion to spread haphazardly over the entire region, particularly along a second line joining Peretola to Osmannoro to Campi Bisenzio and a third southern line connecting Scandicci to Signa to Lastra a Signa.[9] In 1964, the textile industry demanded concessions for its development, inducing the municipal administration of Prato to allow industry to spread right up to the limits of the alluvial plain (the area between the Bisenzio and Arno Rivers) to occupy any area where groundwater was obtainable. In this way the entire area became one single manufacturing district, characterized by the widespread presence of small to medium-size factories.[10]

In the course of this second phase the workforce rose from 90,235 to 101,539: the manufacturing industry continued to increase (by 84.64%), whereas industries connected with agriculture and quarrying were shrinking. The construction industry progressively increased in size: between 1951 and 1981 the number of dwellings in the Province of Florence rose from 210,980 to 410,299 (+98.26%) and the number of rooms from 971,849 to 1,873,536 (+92.78%).[11]

ENVIRONMENTAL EMERGENCIES

From the second half of the 1950s the environmental impacts of urbanization and industrialization have been so considerable as to constitute a true ecological crisis. The most pressing emergency concerned the water cycle. Around the mid-1950s health inspectors, requested by the province's health official, began issuing alarming reports on the state of the Arno's tributaries. Of particular concern were the discharges from textile plants as well as domestic effluents, which in the absence of adequate sewerage infrastructure were spilling directly into the water bodies. The situation appeared difficult to tackle right from the start because industrial plants and craftsmen's laboratories were scattered over a vast area, often within residential areas. This made the creation of an efficient sewerage system very difficult (Florence's sewers were still those built between the 1860s and 1870s). Even more problematic was finding a site to set up a purification plant; Florence did not possess adequate council property on which to build such plant. Entrepreneurs, although aware of the need to treat wastewater, were not prepared to bear the cost of purification.

To compound the situation, around the mid-1960s water bodies in the entire Florence-Prato area were, de facto, biologically dead,[12] which contributed to the seasonal character of the Arno and of its tributaries. During the summer months the mean volume of the water carried by the river and its tributaries decreased considerably, causing serious problems for water treatment plants located along the river upstream (Anconella) and downstream (Mantignano) from Florence.

This caused a twofold conflict about the use of water resources. The first pitted Prato against the municipalities of the Florentine area—faced as they were with a chronic shortage of drinking water due to considerable population increase—and against Florence itself, whose administrators continually delayed the creation of a consortium for the collective management of water supplies. This delaying tactic stemmed from the city's refusal to finance the entire operation, as the smaller towns requested.

The second conflict concerned the industry associations (Prato's in particular) and town councils. The significant depression of the water table prevented any increased pumping from underground wells, which constituted the primary source of water for factories. Although they had been unwilling, because of the increasing costs involved,[13] factories were then compelled to resort to the surface waters also used by municipal utilities to feed their waterworks. Hence their urgent requests to devolve an ever-increasing quantity of Arno and Bisenzio water to industrial needs.

Pollution and scarcity of water resources were not the only causes of the ecological crisis. Since the second half of the 1960s the district of Florence had some of the most polluted air. Industrial plants were voracious energy consumers (about 17% of the regional demand), which derived mostly from oil products (76%), but also from methane gas (12%) and electricity (11%). The high consumption of energy accounted for significant amounts of noxious emissions, so much so that by the end of the 1970s the communes of Florence, Calenzano, Campi Bisenzio, Scandicci, Signa, and Fiesole were included within the control zone "A" as laid out by Law No. 615 of 1966.[14] If we add to industrial emissions the fumes of domestic central heating and vehicle exhaust, it will not be difficult to imagine the quality of the air in the city, which contained high percentages of sulfur dioxide, robust doses of lead and polycyclic aromatic hydrocarbons, traces of minerals (silica, magnesium, copper, zinc, nickel, manganese, calcium, and vanadium), and malodorous substances (mostly phenols and fatty acids emitted by food processers and tanneries).

The third emergency concerned the disposal of solid waste. By the end of the 1960s the constant increase in industrial production and accelerated rates of private consumption created an evident problem. The main problem occurred in Florence, which was unable to get rid of its household waste with only two civic dumping sites at Mantignano (where, incidentally, the second inlet to the aqueduct was located). Beginning in 1966 Florence had to export its urban waste (and along with that its ecological crisis) well beyond its borders, first into the region of San Piero a Sieve (Mugello) and from 1971 into Certaldo (Valdelsa). Over the years Florentine city councils had to face numerous conflicts with those communes whose citizens suffered the problems created by the dumps. During these periods of crisis, the Municipal Department for the Disposal of Urban Waste had to place the waste on its own premises (which was within a residential district of the urban area) or in other, provisional sites. In the early 1970s a case of unauthorized dumping in the Santerno valley (Imola) elicited harsh reactions from the councils of both Ravenna and Bologna. The situation found a provisional solution with the opening of the San Donnino incinerator in 1973.

Finally, we may regard the swift progress of the hydrogeological degradation as a fourth emergency. The principal agents of soil depletion were the quarrying (numerous quarries, often unlicensed, were scattered along the bed of the Arno) and construction industries. Beginning in 1951 these industries started a progressive invasion of the riverbed and of those areas adjacent to the river which served as natural expansion basins in case of exceptional floods; these areas were destined for industry or housing.

CITIZENS AND THE ENVIRONMENTAL CRISIS

Between 1946 and the early 1970s ecological issues achieved only limited consensus among the people, attracted as they were by the idea of coming out of a condition of mere survival and by the possibility of reaching levels of consumption hitherto enjoyed only by the most affluent. Consensus about a swift shift to industrialization was almost total, and questioning of economic development was strongly opposed.

For this reason, notwithstanding the incipient deterioration of the urban environment, no organized opposition emerged. There were only reports to the Provincial Health Office (the only full-time authority in charge of monitoring the environment, under the National Health Service) regarding blatant episodes of pollution, which were made by individuals or small groups of citizens annoyed by the fumes or by the effluents issued by industrial plants situated next to their homes. The objects of these complaints were episodes of water and air pollution and also the difficult conditions in the new suburbs. Complaints came from citizens exasperated by the smells issuing from numerous small streams crossing the Florence-Prato area (many of these, directly receiving domestic sewage and industrial wastewater, had become true open sewers) and from sport fishing associations alarmed by the shrinking fish populations. There were also reports from health inspectors who highlighted the fact that riverbanks had become dumps where any kind of waste (often including industrial waste) was discarded.

The presence of factories within densely populated urban districts fueled numerous protests from those who lived in the vicinity. In the course of inspections requested by the provincial health officer, it often happened that the municipal police had to intervene in quarrels between citizens annoyed by industrial fumes and workers who did not welcome health inspections for fear of a possible closure (however temporary) of their workplace. Residents lodged numerous complaints about their living conditions in unhealthy and cramped dwellings located in the new suburbs, which lacked even the most basic sanitation, public greenspaces, and places for socialization.

THE COMING OF AWARENESS, 1971–1982

THE FIRST MEASURES

By the beginning of the 1970s the environmental crisis continued to manifest itself mainly in the water supply. The most apparent problem was surface water pollution, highlighted by a whitish froth floating on the Arno and by the extravagant colors acquired by the streams where textile plants discharged their effluents. In an attempt to halt such pollution, the Florence Town Council passed an ordinance banning the use of nonbiodegradable detergents, to go into effect on January 20, 1971. As Mayor Luciano Bausi stated at a meeting called to explain the initiative, the measure had become necessary in an attempt to halt water contamination with anionic surfactant chemicals, which had reached levels much higher than those set by the World Health Organization (WHO; 500 μg/liter); tests carried out in December 1970 by the District Hygiene and Prophylaxis Office had shown values ranging between 1,900 and 3,800 μg/liter.[15] Although its significance was little more than symbolic, since it ignored the wastewater containing chemicals (chromium and cyanides) even more toxic than the detergents dumped into the Arno by factories, the Florentine measure prompted a debate that acquired a national dimension, causing passage of a bill which addressed the "biodegradability of synthetic detergents," presented by the Health Minister, Mr. Mariotti.[16]

In 1972 the regional council passed a resolution in which it invited the communes to unify quality standards for "domestic and industrial liquid effluents," along with those suggested by the Regione Toscana.[17]

Two years later the first *Water Pollution Map* was published, and a regional law (No. 22, May 27, 1974) was passed which provided funds for a program of "works for the search for and utilization of water resources and for the disposal and purification of wastewater."[18] There were measures essentially aimed at solving the shortage of potable water—regarded as a priority since the Florence's aqueduct was fed mostly by the Arno—which could not cope with the crisis of the entire Florentine water system, afflicted as it was by two more problems now turned endemic: a scarcity of water resources and a serious hydrogeological upheaval, revealed in all its gravity by the flood of November 1966. Notwithstanding the flood, during the first half of the 1970s the political debate still languished around the proposals put forward by the Interministerial Commission for the Study of Water and Soil Management (Commissione Interministeriale per lo Studio della Sistemazione Idraulica e della Difesa del Suolo), jointly instituted in 1967 by the ministries of Public Works and Agriculture. In order to protect Florence from new disastrous floods, the commission had proposed a system of twenty-three reservoirs whose role would be to cut the flood crest before it reached the city center. In 1970 this project was heavily criticized during a meeting called "A Plan for the Arno" because, according to its critics, it confronted the problem of defense against floods in "a most traditional and narrow-minded manner, in the way of a true *war against water*" without contemplating an assessment of future needs for drinking, irrigation, and industrial water requirements, or a balanced distribution of available resources.[19]

The situation came to an impasse difficult to overcome, first for a latent conflict between the regional administrations, which since their creation appeared prepared to face the environmental crisis, and the communes, much more lukewarm toward ecological problems. The administrations of the area's industrialized communes were staunchly opposed to any measure restricting the expansion of residential and industrial developments; the haphazard urban and industrial development was tolerated since administrations believed it would alleviate social tensions generated by unemployment and the lack of housing. In the second instance, pressures exercised by industrialists who—notwithstanding their apparent support in principle for projects aimed at controlling the Arno basin—showed firm opposition to any revision of urban planning and were scarcely inclined to foot the bill for an environmental cleanup program.[20]

Finally, a "Scheme for a Pilot Project for the Control of the Arno basin" (Schema di progetto pilota per la sistemazione del bacino dell'Arno) was passed; the document envisaged protection of the city of Florence from floods and the monitoring of the Arno River according to a "coordinated program for the use of water and the control of its quality."[21] Five years later, in 1978, the final text for the Pilot Project was presented, which envisaged the "building of reservoirs for mixed use (to regulate and contain floods) or for the sole purpose of regulating the water flow," the "realization of defenses against floods," and the construction of treatment plans aimed at reducing by 90% the pollution caused by wastewater. The document highlighted the chronic lack of drinking water resources and stated that the building of new waterworks was "a matter of utmost urgency" and "prejudicial to any development program." It established two priorities for the industrial uses of water: ensuring supplies for factories—those of the Prato textile industry in particular—and making the process of industrialization compatible with available water resources.[22] Notwithstanding such a disconcerting picture, actions, even those regarded as priorities, remained at the planning stage. Among these were a basin for the prevention of floods, two more reservoirs destined to feed the waterworks, an outlet canal for the Arno, and four purification plants for the Florence-Prato area.

In 1980, in observance of measures established by Laws No. 319/1976 and No. 650/1979, the Regione Toscana approved the first biennial program (1980–1982) of the *Regional Plan for the Purification of Waters* (*Piano regionale di risanamento delle acque*). This document too was little more than a declaration of intent which reiterated interventions already included in the Pilot Project. Concerning the Florence-Prato area, the document spoke of "intervention schemes at a considerable level of definition and depth" listing a series of actions all in the early stages of planning.[23]

Twelve years of discussions had produced a significant number of plans for the water system, none of which ever came to fruition. The realization that a serious environmental crisis existed led only to the approval of a number of sectoral policies aimed at ensuring the availability of resources in order not to hamper economic development and to alleviate the more urgent negative effects (water and air quality, chaotic urban development). The main initiative undertaken by local authorities to reduce air pollution was the creation of a consortium for the distribution of methane gas, whose objective was to encourage the shift from oil to gas for central heating systems. By the end of the

1970s the area south of Florence still did not have a methane gas supply, and methane gas pipelines did not reach any of the other communes.

Choices in matters of public and residential housing remained based upon a policy of alterations to the town plan without strategy. The result was a failure to achieve a concerted strategy involving all communes—the comprehensive Florentine urban area plans of 1973 and 1978 were nullified by the attitude of local authorities, who continued to ignore the guidelines contained in the regional documents. The doubling of the urban perimeter of Florence resulted from vast public housing developments on the far outskirts of the city, as well as from the sale by apportionment of the hills, which were then densely built up.

THE EARLY STEPS OF THE ENVIRONMENTAL MOVEMENT

During the second half of the 1970s the environmental movement began to take its first faltering steps. The main organizations (those most firmly rooted in the Florentine-Prato area were Italia Nostra, the World Wildlife Fund [WWF], and Pro-Natura) still privileged the conservationist approach: at the core of their actions was the conservation of the landscape (woodlands and mountains in particular), the protection of fauna, and the persuasion of public opinion toward the establishment of protected areas. In the urban areas environmental groups focused their attention primarily on water resources (e.g., pollution and shortages caused by an irrational use of water) and on some urban planning issues (e.g., building without permits and conservation of historical centers).

We cannot yet speak of an environmental movement capable of influencing politics because the various organizations were not capable of mobilizing a significant number of people prepared to take action in defense of the environment. People were still not very focused on environmental issues because they appeared rather irrelevant compared to their own economic problems (these were the years of reshaping the systems of production, which involved divestments of industrial assets and dismissal of workers) and those of public order (the 1970s are identified with the age of terrorism).

INTERVENTIONS FOR RESTORATION, 1982–2001

PERSISTING PROBLEMS

In a book published in 1987, the commune of Florence described the area's environmental problems as "evident" and "of considerable gravity."[24] The most pressing crisis continued to concern water resources. Every year, just in the commune of Florence, water consumption amounted to 70 million m^3 (10 m^3 pumped out of underground wells, 5–6 used by factories). The chief problem remained that of water quality because almost all untreated liquid waste was discharged into surface waters: less than 10% of all discharges were purified (industrial wastewater represented around 4–5% of the whole). In summer, especially in August, the flow of the Florentine sewer system was almost equal to that of the Arno (4 m^3/sec), enough to frequently halt the intake of the Mantignano waterworks (west of the city). A very worrying situation was that of the water table, which was becoming polluted with numerous chemicals (organic chlorines in particular).

The second problem was represented by air quality. The main responsibility lay with traffic, which every year discharged into the atmosphere as much as 15,000 tons of carbon dioxide, 1,500 of unburnt hydrocarbons, and 3,500 of nitrogen oxides. Domestic central heating, largely fueled by methane gas (80%), produced around 300 tons of sulfur dioxide. Industrial activities emitted sulfur dioxide, heavy metals (lead, zinc, cadmium, chromium, mercury, and copper), acids, phenols, aldehydes, carbon compounds, sulfur, and nitrogen. Air was also affected by acid rains composed of nitric, sulfuric, and chloridic acids, and heavy metals (lead, zinc, cadmium, mercury, and copper). The most serious consequences were visible in the historic center of Florence, where the surface of marble monuments turned to chalk, and in the forest of Vallombrosa.

Urban planning problems also remained unsolved. The city's study, *Firenzecologia*, assessed Florence as having "reached saturation" for housing and that the outskirts of the city had expanded "beyond measure" with housing. In view of this it recommended a transformation of the city from monocentric into bipolar, in order to lighten "greatly the urban pressure on the historic center."[25]

A further problem concerned urban solid waste. Already in 1978 the production of solid waste (160,000 tons/year) surpassed the capacity (130,000 tons/year) of the incinerator located in San Donnino.[26] The closure of this plant by the provincial administration in 1986 increased the problem of waste disposal, forcing the town council to send urban waste to wherever there was a plant in Italy prepared to accept it.

The picture presented by the town council enables us to highlight the twofold nature of the environmental crisis. On the one hand, this was the direct result of a development model that had concentrated its driving forces on the construction and manufacturing industries, both voracious consumers of resources (land, water, energy). On the other, it largely depended on the rising of living standards and the sharp increase in consumption and pollution caused by lifestyles and consumption habits. The most prominent fact, but not the only one, was the deterioration of air quality due to traffic. Between the end of

Figure 2.3. In the background: the northwestern suburbs and the new courthouse in the Novoli area (2015). *Credit*: Federico Paolini.

the 1960s and the first half of the 1970s, washing machines—which had swiftly supplanted hand-washing with homemade soaps and lye—were to blame for the contamination of water with industrial soaps. The solid waste crisis was a further example. The problem became urgent in the first years of the 1980s when mass distribution, encountering a wide acceptance by consumers, contributed to the establishment of a throwaway culture, the consequence of which was the rapid growth in solid waste, composed of 31% paper and plastic, the chief materials of the new style in consumption.

THE HARD WAY TO CREATE AN ENVIRONMENTAL INFRASTRUCTURE NETWORK

During the first half of the 1980s environmental policies were characterized by significant activism in the region—foremost concerning the cleansing of the Arno,[27] regional planning,[28] and energy[29]—which was opposed by trade corporations, trade unions, and communes, causing de facto a continual postponement of the enforcement of regional directives. This resulted in further delays to the updating of civil and industrial purification plans (according to the terms established by the Merli Law of 1976) to June 13, 1989. The communes adopted only a minimum of the regulations for urban planning, Tuscany's energy deficit remained high (18.7%), and energy consumption continued to rely on the dual sources of oil and methane.

This new impasse had been caused by pressures exerted on local authorities by small and medium-size industries in particular and by trade unions. The former demanded significant contributions as a condition for adopting antipollution technologies. The latter, while denouncing environmental impact problems, placed energy deficits and the protection of jobs at the core of their worries.[30] In a period during which serious restructuring of companies was in progress, the district councils of the industrialized communes of the Florentine and Prato region postponed adoption of concrete environmental measures, which large numbers of citizens perceived as an obstacle to economic development.

In the early 1990s, when the need to improve the quality of the urban environment became urgent, local authorities approved some measures aimed at ensuring acceptable air and water quality levels by means of new infrastructure: ten new plants for the purification and treatment of civil and industrial effluents became available (the capacity of which, however, did not suffice for more than eighty-eight thousand residents), a new air quality monitoring network; and a provincial plan for the disposal of urban and industrial waste. In the Prato area a special waterworks for recycling water was built, and the construction of a dam at Bilancino (Mugello)[31] was begun in order to resolve the perennial shortage of water afflicting the Florence area in summer (though it was completed in 1995, the reservoir's operation was delayed until 2002). These efforts mitigated the most blatant effects of water pollution, and the levels of atmospheric pollution were controlled thanks to the introduction of methane gas, lead-free gasoline, and catalytic converters.

CHAPTER 3

FLORENCE AND ITS WATERS

WATER MANAGEMENT POLICIES IN A CENTRAL ITALIAN CITY, 1944–1980

This essay illustrates a synthesis of research[1] conducted on the environmental history of the relationship between a city (Florence) and its water resources. The purpose of this chapter—which offers the first results of a much larger study dedicated to the environmental impact of urbanization and industrialization in the greater Florence area from 1946 to 2005[2]—is to offer some thoughts useful to the international historiography of water and, in particular, of problems relating to urban development and water management policies.[3]

BETWEEN RECONSTRUCTION AND DEVELOPMENT, 1944–1965

WATER MANAGEMENT POLICIES

The city of Florence has since time immemorial shared its space with rivers. The largest is the Arno—the eighth river in Italy in length and importance (241 km)—which divides the city in two, crossing it from east to west. From the surrounding hills the torrents Mugnone, Terzolle, Mensola, and Affrico (the last channeled underground in its lower reaches) descend into the Arno. There also are a number of lesser streams (Santa Cristina, Santo Gervasio, Piagentina, Anconella, and Ricorboli) that have been channeled as a consequence of urban development. The western part of the city is also crossed by the Fosso Macinante stream, whose original function (already mentioned in documents from the first half of the fourteenth century) was to collect the waters of the

Arno in case of flooding and divert them toward the open countryside, while in the mid-sixteenth century it powered a series of mills. In the course of time, Florence, situated as it is in the middle of a basin surrounded by mountains, has suffered from numerous floods, often with disastrous consequences, the most serious of which, before the 1966 flood, took place in 1269, 1333, 1547, and 1844.[4]

These frequent inundations have always been enhanced by a profusion of destabilizing geological events which characterize the Arno basin and which are due to the nature of the soil: impermeable and prone to erosion. Beginning in the first half of the twentieth century, to these natural disruptions have been added a pattern of growth based on urban expansion and industrialization, which has deeply altered the use of resources.[5] Florence and the communes of its province are typical of how a region can be deeply altered by technological and social factors (e.g., Fordist industrialization, urban expansion, intensive use of oil and electricity, and the expansion of private transportation), which J. R. McNeill identified as the "cluster" of the motorized city.[6]

The consequence of the acceleration of urbanization and industrialization, which has taken place since 1950, has been a radical transformation and a serious deterioration of the environment. The most significant effect has been a rapid geological disarray caused by the neglect of upland and hill agriculture, leading to the collapse of the capillary drainage system of cultivated fields; by the dredging of riverbeds to obtain building materials; by a significant increase in earthworks (e.g., excavations, trenches, ditches), indispensable for creating infrastructure, which led to soil degeneration; by the progressive occupation of riverbeds—of those portions of riverbeds which diverted the overflowing river—by residential and industrial buildings.

Between 1944 and 1965 six municipal administrations and two prefectural commissioners[7] governed the city, but neither was able to work out coherent policies for governance of the region and the environment. First, the dramatic budget deficit made impossible the enforcement of necessary measures for confronting the municipality's numerous problems.[8] Second, the construction industry was placed at the economic center of the region. In fact, the principal political and economic actors regarded urban development in the area northwest of Florence, toward the municipalities of Sesto Fiorentino and Prato, as the greatest opportunity for the growth of the entire Florentine economy (Figure 3.1). This made city planning disliked by vast sectors of Florentine society, which regarded the problem of regional organization secondary to the lack of housing and the high unemployment rate.[9] During the mandate of the third La Pira administration, the town plan promoted by socialist councillor for Urban Planning Edoardo Detti[10] was opposed by La Pira himself,[11] by the con-

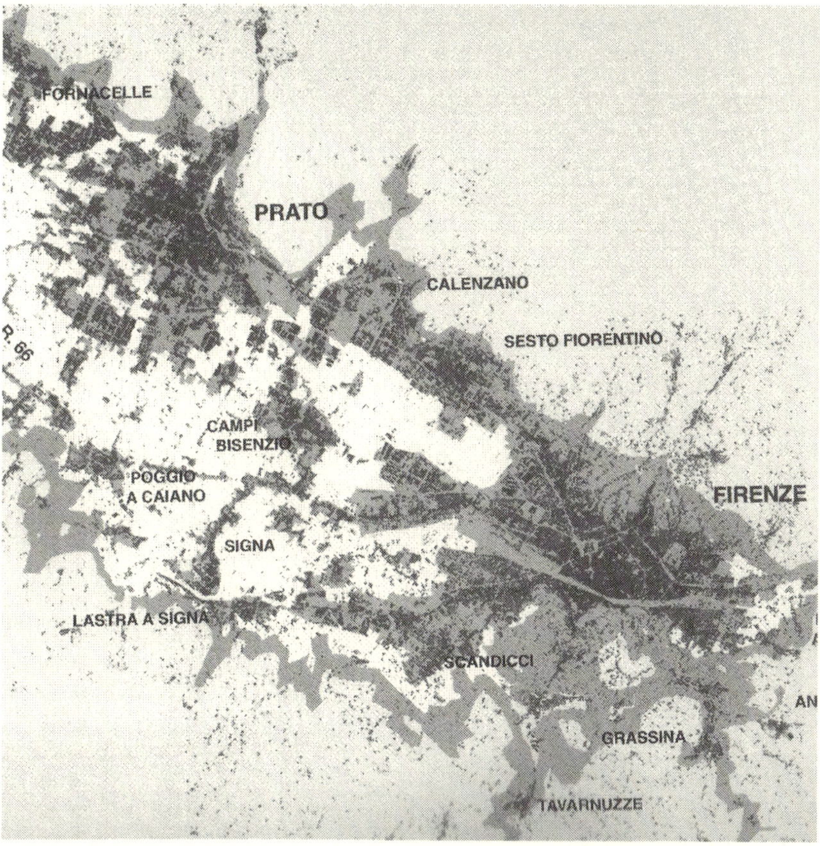

Figure 3.1. A map of Florence and the Florentine area. *Credit*: Margherita Azzari (ed.), *Atlante geoambientale della Toscana* (Novara: Regione Toscana e Istituto Geografico DeAgostini, 2006), 90.

struction workers' unions (worried about possible repercussions on employment),[12] and by industrialists.[13] In 1965, in the aftermath of the crisis which led to the fall of the La Pira administration, the town plan was dropped.

In this situation the good management of water resources and land came second to the needs of urban expansion. The main actions were a study for a river management plan contemplated by a law proposed in 1952 but which never came into being, and a study of the geological situation of the Arno basin, promoted by councillors for agriculture in the provinces of Tuscany in May 1965. It was not even considered that the existing flood warning system, consisting of antiquated hydrometers and pluviometers totally disconnected from each other and without coordination, might be improved to monitor the entire Arno basin.

Florence's water management policy was chiefly directed toward sewerage and water supply systems. Until the 1950s the drainage system of Florence was still the same one created in the 1860s and 1870s, consisting of three outlets (central, northern, and southern) of the manifold drain called Goricina, which flowed into the Fosso Macinante (which itself became a manifold drain), and of still more ancient drains (i.e., the so-called "great drains of Ripoli and Gusciana").[14] The need to modernize the entire sewerage system in answer to the requirements of an ever-expanding city was discussed in a meeting of the city council on November 6, 1950, when, for the first time, a proposal was made to put into operation a combined sewerage system which had already been approved by a special Health Authority Commission in the 1940s. It was envisaged that the scheme, which had no budget, would be completed within ten to fifteen years.

The problem of the drainage system became an issue again only in 1957 when the city council contemplated the building of a new residential quarter with six thousand apartments in an area called Sòrgane southeast of the city. It was necessary to provide the area with a suitable sewerage system capable of servicing the new development. Between the end of the 1950s and the first half of the 1960s, however, successive administrations renounced a project of general restoration and restricted themselves to repairing the old drains wherever necessary. In March 1964 the city council approved a statement denouncing the faults of the city's sewerage system. In the following years—during which the city was governed by a prefectural commissioner as a result of a complex political crisis which the parties were unable to resolve—this denunciation fell into oblivion, until the great flood of 1966 brought home the structural faults of the entire drainage system.

The municipal waterworks was not in better condition.[15] From the time of the city's founding to the first decade of the twentieth century, the chief supply for Florence had been underground water. Almost every house had its own private well (more than ten thousand). The city's waterworks was itself relying on the water table until 1912, when it was decided that the waters of the Arno should be exploited by means of a plant to be built in a place called Anconella in the southern part of the city. In 1928, during the Fascist era, the Podestà administration authorized the building of a municipal water pipeline from S. Maria a Mantignano.

By 1946, soon after the end of the Second World War, there were essentially three problems with the municipal waterworks: (1) the inadequate capacity of the water pipeline (carrying about 60,000 m^3 per day, while the city needed 85,000 m^3); (2) recurrent water crises because of the seasonal nature of the

Arno, which caused insufficient water flows in summer (below 3 m^3/second), resulting in underutilization of the Anconella plant; and (3) the precarious hygienic situation of the water pipelines, which almost entirely ran inside the drains.

Not until 1953 did the municipal administration charge a commission with the task of finding a way to increase the production of drinking water. The final report, produced in 1954, recommended the Florentine administration join with the electricity company Selt-Valdarno to exploit the reservoirs of Levane and La Penna and the torrents Carza, Carzola, and Terzollina (situated in the Bisenzio River basin), and the creation of a reservoir by damming the Bisenzio itself (a river which crosses the city of Prato after descending from the Tuscan Aemilian Apennines), both for domestic use and for the production of electricity. The proposal to build a dam along the course of the Bisenzio caused a conflict between Florence and Prato, whose municipal administration in April 1956 passed a bill stating the "absolute binding necessity, for the city of Prato and for its region, to use all the water of the Bisenzio."[16] Dissent erupted again in 1957; helping to fuel the clash came the proposal, presented by a study group led by the engineer Giulio Supino, to harness the waters of the Bisenzio in order to reserve the water table for industry and sell part of its excess to Florence. The proposal caused protests among industrialists and the Christian Democrat Party of Prato, which in a meeting reiterated Prato's exclusive rights to the waters of the Bisenzio, the only river in the Prato administrative area.

Consequently, in November 1958 the Florence City Council approved a plan to confront the perennial water scarcity—in order to increase the production of 2,000 liters per second—by building two new waterworks: a reservoir supplied by the torrents Carza and Carzola, and damming the torrent Pesa, with a capacity of 7 and 17 million m^3, respectively. In 1958 a proposal was presented by the civil engineer Cambi for the creation of a reservoir on the Sieve River, which was predicted to supply the waterworks of both Florence and Prato. The plan had the effect only of stopping the building of the other two plants, causing a delay that lasted nearly three decades.

In 1961, meanwhile, a commission (known as "the Seven") was allotted the task of presenting the municipal council with the best possible alternative to a new waterworks. The primary choice was damming the Sieve and, secondarily, exploitating the torrent Ema. The job of developing the executive project for the Sieve Reservoir (Lago di Bilancino) was not given to the man who proposed it, the civil engineer Guadagni, until June 1963. Two more years passed before the administration of Florence finally approved, in December

1965, a resolution to begin a study of the new city water supply system, which now included the Sieve Reservoir and waterworks. On the eve of the November 1966 flood, the hydraulic problems of Florence remained unsolved: the city's network of sewers and water mains was the same one put into operation between the nineteenth century and the 1930s.

POLLUTION

From the beginning of the 1960s the problems of pollution added to infrastructural deficiency.[17] The water quality of rivers and torrents became a concern, not only in Florence but for the entire metropolitan area.[18]

The water quality of the Arno was beginning to deteriorate on the eastern outskirts of Florence where the river received the wastewaters of the Compiobbi (whose raw sewage contained residues from galvanic processing plants and synthetic rubber factories) and Bagno a Ripoli suburbs (whose raw sewage contained residues from paper mills and chemical plants). At the extreme western outskirts Arno's water quality deteriorated further because its tributaries, the Mugnone and Terzolle, were highly polluted by raw sewage, the drain waters of the factories in Rifredi, and the wastewater of the Careggi Hospital. After having received the waters of the Greve, Vingone, and Bisenzio, the water quality of the Arno near Signa became critical. Coming out of Florence the river carried significant levels of ammonia, nitrites, nitrates, surfactants, phosphates, chlorides, and traces of chromium and cyanides.[19] The water quality of the tributaries was actually worse than the Arno itself, with the exceptions of the Sieve and the Pesa, thanks to the scarcity of industrial plants and urban conglomerates along their courses. The Greve appeared highly polluted by sewage and surfactants, which reached "enormous values" and were responsible for the pollution of a large number of wells.[20] The torrent Vingone collected and discharged into the Arno the sewage of two of the most densely industrialized areas of the province, Scandicci and Lastra a Signa, housing foundries, galvanic processing plants, silver processing plants, rubber factories, metalworks, candle factories, soap works, and salami factories.[21]

The Bisenzio carried a high load of pollutants: the sewage of the Vernio, Vaiano, and Prato; the wastewaters of the numerous industries of the Calenzano, Sesto Fiorentino, Campi Bisenzio, and Signa areas; and finally the urban and industrial effluents of the area to the northwest of Florence. The waters of the river contained a high percentage of ammonia, nitrites, surfactants, sulfates, chlorides, and phosphates. Furthermore, the fragments of wool fibers dumped by textile plants (about 4,800 kg per day) caused serious oxygen depletion (each mg of fiber destroys 1.19 mg of oxygen).[22]

Figure 3.2. The areas of the Florentine district flooded by the Arno River. *Credit*: Provincia di Firenze, *Carta dell'alluvione del 1966 nella provincia di Firenze: Pianta dell'abitato fiorentino alluvionato e aree invase dalle acque* (Florence, 2006).

THE BIENNIUM OF THE FLOOD, 1966–1967

On November 4, 1966, the front page of the daily paper *La Nazione* carried an alarming headline: "Dramatic Situation at 6 A.M: The Arno Floods Florence."

Actually, few Florentines would read the newspaper that morning. At 7:20 a.m. some areas of the city had already been flooded, and at 9:35 a.m. the level of the water had risen above two meters in the center of the city, flooding even the Piazza del Duomo. At 7:30 p.m. the RAI (Italian broadcasting corporation) announced that the level of the water was decreasing after having reached 4 meters above street level in many quarters of the city (in some areas of the center it had reached 5 meters).

The flood caused serious damage to dwellings (13,493 families had to leave their homes), to industrial plants (more than twelve thousand jobs were lost, and 3,997 workers in the crafts and trade industries were temporarily laid off or fired), to tourist accommodations (47% of the hotels were damaged), to artistic heritage (over 1,400 works of art, and almost 2 million books in the city's libraries were damaged or destroyed). During the months which followed the main flood, there was a succession of more floods caused by breaks in sewer pipes, while the floodwater—saturated with gasoline and salts (nitrates of potassium and sodium)—continued to cause damage to monuments and dwellings.

During the entire course of 1967 there were discussions on who was to blame for the flood. The first to have the finger pointed at them were the successive local governments of the second postwar period who had failed to build an efficient sewerage and water supply system or a system of flood control for the Arno and its tributaries. There is no doubt that the inadequacy of the sewerage system contributed to the seriousness of the calamity: the flood—which rushed through the city at a speed of 60 km per hour, carrying 4,200 m^3 of water per second—literally caused the explosion of the drains. Then there was the question of the lack of warning which involved the chief of police and the prefect; in a conversation which took place in the early hours of November 4, the two authorities decided not to warn the entire city since the danger "did not seem to justify alarming the general public and for fear of disorder caused by traffic jams as people might rush to reach the highway and the hills."[23] The debate addressed the role of the La Penna and Levane dams (built in the 1950s by the electricity company Selt-Valdarno for the production of hydroelectric energy) situated in Arezzo Province about 50 km from Florence. According to a comprehensive literature, the flood crest which reached Florence originated from the combination of flood waters of the Sieve and of those two dams (about 2,000 m^3/sec from 3 to 6 a.m. on November 4).[24] The argument became animated when the magistrates discovered in February 1967 that Selt-Valdarno had built the Levane hydroelectric plant without the necessary authorizations.[25] However, an inquiry ordered by the public prosecutor's office highlighted the bad conditions of all riverbanks at the time of the flood, thus exonerating the Levane and La Penna dams, which experts reported caused the level of the Arno to rise by only a few centimeters. During the fiery but sterile debate, one of the few useful comments came from Professor Livio Zoli, who held the chair of Forestry and Water Studies at the Faculty of Agricultural Sciences of the University of Florence. He proposed an articulate plan for safeguarding the Arno's banks based on the elimination of the two weirs built across it within Florence (this however would have meant "rebuilding the foundations of all the bridges, as well as raising all parapets and walls of the Lungarni"); the construction of at least ten reservoirs, "with the exclusive role of defense," along the course of the river; and the excavation of an underground canal capable of carrying at least 1,000 m^3/sec beneath the urban area.[26]

Institutions entered the debate with a disorderly array of initiatives, among them a long succession of round tables and a singular proposal: the application of Florence to host the Olympic Games of 1976 (put forward by *La Nazione* for the purpose of accelerating the restoration of the city), which fortunately was later discarded.

In November 1966 the Regional Union of the Provinces of Tuscany issued a statement pointing out "the inefficiency of the civil defense" and proposing the "creation of an adequate system of flood prevention and warning." They hoped for an early "initiation of an effective policy of protection against land degradation." According to the Regional Union, the causes of the hydrogeological hazard were "the indiscriminate excavations of the Arno riverbed" and the "meteoric disruption of the soil and subsoil" caused by a "lack of effective drainage works both in forested and agricultural areas."[27] In March 1967 the Regional Union promoted a round table for discussing a plan for a "system for the rational management of water resources and river basins." In September the mayors of Fiesole, Pontassieve, Rufina, and Pelago invited the municipal administration of Florence to assemble all the communes affected by the flood for the purpose of "avoiding [individual] municipal solutions" in favor of a regional one. In October 1967 representatives of the affected communes of Florence Province met and decided to rapidly assess the damages suffered by water defenses all over the province, to census the repair works already planned and financed, and to monitor works in progress. At the close of 1967 the lack of decision making displayed by the Florence city council[28] and the other provinces of Tuscany (they engaged in calling meetings and conferences which never produced a single plan for the control of river basins) had turned the flood of 1966 into yet another missed opportunity for understanding the causes of soil and river degradation and for starting a unified and definitive program of hydrogeological restoration, placed within the context of a wider regional development plan. Various inquiries point to 1967 as also being the year in which the foundations were laid for a new age of urban expansion in areas subject to flooding risks. The passage of a planning law (No. 765/1967), under the declared intent of helping the economy, allowed the expansion of buildings into all areas of the municipal territory. The result was urban saturation of the Arno basin.[29]

THE UNCERTAIN BEGINNINGS OF RESTORATION, 1968–1980

In the years following the flood,[30] the water crisis became the most evident (and debated) environmental issue in Florence. The most apparent sign was the pollution of surface waters, made visible by an abundance of froth floating on the Arno. In an attempt to halt this degradation, the Florence City Council approved an order which forbade the use of nonbiodegradable detergents, beginning on January 20, 1971. Although it ignored the sewage and industrial waste discharged into the river—which contained materials (e.g., chrome and cyanide) far more toxic than any synthetic detergents—the measure had become necessary in order to contain the contamination of water by surfactants, which had reached levels (1,900–3,800 γ/liter) far higher than those established by the WHO (500 γ/liter).[31]

In 1972 the regional council approved a resolution inviting local administrations to agree on common quality standards concerning "urban and industrial effluents" as established by the regional government. Two years later the first *Map of Water Pollution* was published, and a law was approved (No. 22, May 27, 1974) which provided finances for a program of "works for the detection and utilization of water resources and for the purification and disposal of liquid waste."[32] In 1974 the Consortium of Water Resources Scheme 23 was established, bringing together the communes of Florence, Bagno a Ripoli, Barberino del Mugello, Calenzano, Campi Bisenzio, Florence, Fiesole, Impruneta, Lastra a Signa, Montemurlo, Prato, Scandicci, Sesto Fiorentino, and Vaiano. The constitution for such a consortium ended up blocking action aimed chiefly at the purification of waters, as promoted by the commune of Florence between 1974 and 1975: the construction of a digester (in a hamlet called S. Mauro a Signa astride the communes of Campi Bisenzio and Signa) and two purification plants (in Settignano, a suburb northeast of Florence, and in an area northwest of Florence known as San Donnino). In 1976 Scheme 23 considered building a purification plant capable of servicing a population of seven hundred thousand residents, but it took almost ten years before they agreed on the placement of such a plant, in an area (S. Colombano) astride the communes of Lastra a Signa and Scandicci. In 1980, in observance of Law No. 319/1976,[33] the Regione Toscana approved the first biennial program of the *Regional Plan for Cleaning Up the Waters*. For Florence and the surrounding areas, the document envisaged "intervention schemes at a high level of definition and depth," and listed a series of such interventions, nearly all in their first planning stage.[34] These were measures intended to tackle the issue of water quality, regarded as a priority since the Florentine waterworks was fed chiefly by the Arno. However, they were not adequate to cope with the crisis of the entire Florentine water system, afflicted as it was by two critical factors which by now had become endemic: the serious hydrogeological degradation—as had been exposed by the 1966 flood—and the scarcity of water resources.

To address hydrogeological deterioration, the Ministry of Public Works and Agricultural Policies had instituted in 1976 an Interdepartmental Committee for the Study of Water Management and Soil Defense. Prompted by the eventuality of another catastrophic event like the 1966 flood, the committee worked out a plan (known as *Piano Supino*, from the name of its rapporteur, Professor Giulio Supino) for the reorganization of the Arno basin.[35] Appointed by Florence's College of Engineers, the Interdepartmental Committee recommended the construction of a series of reservoirs of medium (10–50 million

m^3) and smaller (1–10 million m^3) capacity. The lower section of the reservoirs was to be used to accumulate water for the production of electrical power, and also for domestic and irrigation uses. The higher sections should instead have been used to prevent floods. According to the *Piano Supino*, in order to protect Florence from a new flood, seventeen such reservoirs and a drainage channel would be necessary in the tract of the river upstream of Florence in order to allow the riverbed to carry no more than 2,200 m^3 of water per second (the 1966 flood carried 4,200 m^3/sec). Because of the high costs (over 65 billion liras, at 1968 values, for only the period from 1970 to 1975), the Interdepartmental Committee's recommendations were shelved, never becoming operational.

Ten years later, in 1978, the *Final Report* of the *Progetto pilota per la sistemazione del bacino dell'Arno* (Pilot project for the restoration of the Arno Basin) was approved.[36] Compared to the *Piano Supino*, the Pilot Project envisaged the building of eleven reservoirs (two along the course of the Arno and nine along its tributaries) for a total capacity of 378.8 million m^3. Once again, however, due to high costs (about 111 billion liras at 1971 values) and the lack of adequate finances, the planned interventions were suspended.

The availability of water was also an issue. From 1959 to 1969 mean daily water consumption had more than doubled to a mean per capita requirement of 372 liters. In 1970 the Florentine water treatment plants produced 172,000 m^3/day: the Anconella plant—by now technologically inadequate, scarcely safe, and very costly—provided 135,000 m^3, and the Le Cascine plant (fed by twenty-five wells constantly threatened by sewage pollution) provided 25,000 m^3. The Mantignano plant produced only 12,000 m^3 because the constantly lowering groundwater levels had reduced the supply by more than 8,000 m^3/day. The ongoing water table depletion was particularly serious since it prevented any increased exploitation of underground wells, which provided about 20% of the need. The deterioration of water quality also prevented increased exploitation of wells, a significant number of which were polluted by ammonia and organic chlorinates. This meant that in Florence only one source of water remained certain: the Arno.[37] In order to confront the water scarcity afflicting the city in October 1970, the municipal administration of Florence presented an intervention plan to upgrade the Anconella water treatment plant (doubling the production up to 3,000 liters per second within ten years) and to quickly construct the Bilancino Reservoir in conjunction with the Sieve waterworks, which would have ensured a sufficient supply of drinking water until at least 2000. Meanwhile work to separate the water mains from the sewerage system got under way. Notwithstanding the ever-increasing problems caused by water scarcity,[38] by the early 1980s the Florence City Council was still arguing about the opportunity to build the Bilancino Reservoir and a new Sieve aqueduct—after twenty years of constant bickering without having reached a final decision.[39]

* * *

Thirteen years of discussions on water management policies had produced a significant pile of documents which never actually started the process of restoring the water system. In the second half of the 1990s Florence still continued to wait for an efficient water supply system, a purification plant (work on the S. Colombano plant was started on June 18, 1994), and an efficient flood control system (the Bilancino Reservoir was completed in 1995 and became operational in 2002). The continual deferment of water management policies was due to a complex series of factors. First, it was the peculiarity of the process of urban and economic development of Florence (as in other important Italian cities) which, to use Melosi's expression, had caused "the age of miasmas" partly to coincide with the "urban crisis in the age of the ecology."[40] Italy had gone from a socially backward country, with an almost entirely agricultural economy, to that of a modern industrial country with an advanced and complex social and political organization in little more than twenty years, from 1950 to the first half of the 1970s. The modernization process had been very fast and extremely chaotic.[41]

After the Second World War, for example, Florence experienced a migration wave without precedent, made up chiefly of young adults coming from the surrounding countryside, from unindustrialized Tuscan provinces, and from the southern regions in search of jobs in the factories scattered around the municipality. To alleviate the consequent social tension, Florence's municipal administrations began to build working-class neighborhoods in areas unaffected by building speculation. The new developments lacked the most elementary sanitary and other infrastructure, such as sewers and waterworks, causing Florence to be afflicted with a succession of typhoid fever epidemics lasting to the early 1970s. In this context the absolute priority for the local administrations of Florence and its adjacent municipalities was to alleviate social tensions fueled by poverty and a lack of housing.

Because of the scarcity of finances, caused by considerable debt and low state contributions, Florence's municipal administrations were unable to enforce prompt and effective water management policies. The Florence administrations' ambivalence in matters of water management policies was also encouraged by inadequate national legislation: up to the early 1970s, the main national legislation on water management were Law No. 2258 of March 20, 1865 (on public works), ministerial instructions of June 20, 1896 (local rulings on soil and dwelling hygiene), Law No. 523, July 25, 1904 (on hydraulic works), and Articles 217 and 227 of the health law (Royal Decree No. 1265, July 27, 1934). Only in the second half of the 1970s was a law passed (No. 319/1976, see note 13) whose implementation, due on June 13, 1979, suffered a series of procrastinations and was eventually modified by Law No. 650 of December 24, 1979.[42]

There are at least three more causes for delay in the enforcement of water management policies. The first was a conflict between the municipal administrations and the Regional Administration of Tuscany (Regione Toscana). As we have seen, the earliest water management policies began to be discussed, and under political pressure, the Regione Toscana began to implement those policies in 1970, as did other regions. Water management policies promoted by the Regione Toscana encountered opposition from municipal administrations, who regarded them as an impediment to further expansion of residential and industrial developments. Second, the conflict set the communes of the Florentine area (confronting a serious shortage of drinking water resulting from a considerable population increase) against the commune of Florence itself. Florence continued to postpone the creation of a consortium for the collective handling of water management policies because it was not prepared to foot the bill for the entire cost, as was expected by the other communes. The third was the fiery confrontation between the Florentine division of the Confederation of Italian Industry and municipal administrations. The dissatisfaction of entrepreneurial organizations was caused by water scarcity due both to insufficient quantity and the incipient decline in quality. More and more often groundwater turned out to be unfit for industrial use due to pollution and excessive hardness.

Furthermore, the serious subsidence of the water table prevented an increase in pumping from underground wells, which represented the primary source of water for industrial plants. Although unwilling because of its high cost, the plants were forced to resort to surface water hitherto used to feed municipal waterworks. This pressing demand to reserve a growing quantity of Arno and Bisenzio water for industrial needs became the highest in the entire region, 3.11 m^3 per day per worker (this is an approximate but realistic figure translating into a daily requirement of about 419,000 m^3). Such conflicts caused a continuing postponement of the enforcement of water management policies which—notwithstanding discussions dating back to the 1950s—were not put into effect until the second half of the 1990s and the beginning of the twenty-first century.

CHAPTER 4

INDUSTRIES, TOURISM, AND THE ENVIRONMENT

A CASE STUDY OF TUSCANY'S CENTRAL-SOUTHERN COAST

The coasts of Tuscany extend for about 400 km (not including the islands of the Tuscan archipelago). From a morphological point of view the Tuscan coast presents three basic typologies: 1) low coasts characterized by low and sandy shores, with a low gradient and shallow waters even at a considerable distance from the coastline; these are low-energy coasts, with scarce possibility for mixing salt- and freshwater and dispersing pollutants; 2) high coasts with sudden considerable depths; these are high-energy coasts, with a remarkable capacity of mixing and dispersing pollutants; 3) archipelago coasts, with an absolute predominance of high-energy, rocky coasts. The watersheds are crucial for understanding environmental phenomena and problems along the Tuscan shoreline. Six principal rivers discharge a significant pollution load into the sea: an organic load from 10,075,528 residents (population equivalent), 60,418 tons of nitrogen, and 25,504 tons of phosphorus (Table 4.1).

The maximum population density is found in the northern sector (Table 4.2), the stretch from Carrara to Cecina, which represents only 22% of the total length of the Tuscan coast; over 70% of the coastal residents live at a medium density of over 720 residents per km^2 and peaks at over 1,500 in the cities of Viareggio and Livorno. The remaining part of the coast has average densities of a little over 120 residents per km^2.

Table 4.1. Tuscany's major rivers and their pollution load in 2000.

Basin	Organic Load Population Equivalent	Nitrogen Load Ton/Year	Phosphorus Load Ton/Year
Serchio	941,152	3,569	1,164
Arno	7,477,779	34,950	13,496
Cecina	114,169	1,874	959
Cornia	78,110	1,014	478
Bruna	86,791	1,168	547
Ombrone	1,157,308	14,909	7,382
Albegna	220,219	2,934	1,478

Source: Regione Toscana, *La qualità delle acque marine costiere in Toscana* (Florence, 2010).

Table 4.2. Population in the chief coastal communes.

	Communes	Population	Density Inhab./km^2
	Livorno	161,152	1,548
Val di Cornia	Campiglia marittima	12,543	151
	Piombino	33,917	262
	San Vincenzo	6,528	198
Val di Cecina	Bibbona	3,017	46
	Castagneto Carducci	8,204	58
	Cecina	26,355	613
	Rosignano Marittimo	30,558	253
Piana di Scarlino	Scarlino	3,718	42
	Follonica	22,142	397

Source: Regione Toscana, *Piano Regionale di Azione Ambientale 2004–2006* (Florence: Edifir, 2007).

The coastal areas are characterized by two major anthropogenic pressures: industrial activity and tourism. Livorno, Piombino, Rosignano, and Scarlino host industrial areas that are classified as critical with regard to natural resources use.[1] The stretch of coastline from Livorno to Piombino contains most of the heavy industry of Tuscany; the presence of big business makes this area different from that of the rest of Tuscany, which is characterized by light industrial districts and other, scattered industrialization. Much of the iron, steel, chemical, petrochemical, and metallurgic industries in Tuscany are located in Livorno, Rosignano, and Piombino. Livorno contains chemical, petrochemical, and mechanical industries, and a thermal power plant. The Solvay chemical plant in Rosignano produces sodium carbonate, caustic soda, and polyethylene. In the nineteenth century a steel plant was built in Piombino for the production of steel, steel sheets, and welded tubes. Along the Gulf of Follonica there are two chemical plants for the production of titanium dioxide and sulfuric acid.

The municipalities of the center-south coast boast significant annual tourist trade, with an average number of visitors ranging between three and eight hundred thousand. Where major industry has left free stretches of coastline, very intensive summer tourism has developed, causing huge changes to the coast's morphology and generating intense human pressures often beyond the carrying capacity of the area, thus contributing to the exacerbation of some serious environmental problems. Such seaside tourism is concentrated both in time and space, generating congestion on the coast characterized by an irrational use of land and resources. The excessive human presence being restricted to the summer months generates repercussions not only on the natural environment but also upsets the organization of services, which cannot meet the needs of so many tourists. An increased consumption of water, the disposal of waste (in coastal municipalities such demands double during the summer months), and the growing traffic highlight the gravity of the problem of a congested coastline and its consequent deterioration, caused in part by irrational regional planning.

A CONTROVERSIAL DEVELOPMENT MODEL: HEAVY INDUSTRY AND TOURISM

INDUSTRIAL DEVELOPMENT

The existence of port infrastructure, the proximity of mining sites, and the presence of a railway line were conducive to the establishment of industry along the coast. The industrialization process (starting in the second half of the nineteenth century) intensified in the early decades of the twentieth century and then accelerated in the 1950s and 1970s—the period known as the "economic miracle." The process of industrialization prompted a significant repopulation of the coast. In 1861, 10% of the population resided in coastal municipalities; in 1901, 11.22%; in 1951, 20.33%; and 23.29% in 1981.

In Livorno the first factories were built in the second half of the nineteenth century, the main businesses being connected with shipbuilding, metallurgy, and chemicals (glass).[2] In the 1930s a large factory (the Moto Fides Company) produced torpedoes for the Italian Royal Navy, and the first plant of the Livorno Refinery was built. The refinery, completed in 1939, had an annual processing capacity of 1.6 million tons of crude oil and produced fuels (including liquid gas), lubricants, and paraffin. In 1954 the refinery reached an annual capacity of 3.2 million tons of crude oil, destined for the production of fuels and lubricants. The increasing commercial demand for oil products promoted the expansion of the refinery, which in 1971 reached a processing capacity of

5.2 million tons of crude oil per year. Currently, it comprises the main industrial activity in the Livorno area, in addition to some chemical factories (among them Dow Italia, the Italian associate of the Dow Chemical Company) and a thermal power plant. The Solvay plant was established in Rosignano in 1916, starting the production of caustic soda. In 1957 the Solvay Group also decided to start construction of a plant for the manufacture of polyethylene, which required the company to build a pier to enable it to receive liquid ethylene at the low temperatures necessary for the production of polyethylene. In 1959 Solvay decided to start the production of hydrogen peroxide and sodium perborate and in 1971 added the production of plastics. Currently the Solvay plant of Rosignano contains the following production units: *sodiera* (production of sodium carbonate, sodium bicarbonate, calcium chloride), electrolysis (the production of chlorine, hydrogen, caustic soda), polyethylene (resins production); peroxidized products (sodium peroxide, sodium perborate), and chlorinated products.[3]

In Piombino an iron and steel industry was installed in 1865. After a complex series of events which are beyond the scope our narrative, the companies now in the area are Lucchini-Severstal (a full-cycle steel factory, the main Italian center for the production of long sheets), the Magona d'Italia (an engineering company, among the main producers of galvanized and painted metal sheets), Dalmine (galvanized pipes), and thermal power plants administered by Ise Spa (which manages the central services located inside the Lucchini-Severstal plant) and Enel (an oil-fueled power plant with a capacity of 1,280 MW).[4]

A very special case is represented by the plain of Scarlino.[5] During the construction of Scarlino's industrial area, part of the *padule di Scarlino* (Scarlino Marshes) were filled in. These marshes constituted the last remaining example of such ecosystems typical of the southern Tuscan coast. In the early 1970s Montedison (a major industrial group, chiefly active in the chemical sector) completed the construction of a titanium dioxide plant. The local population had welcomed the plant, since it was expected to provide around four hundred jobs. However, industrial production was obstructed at the outset by local governments and by the Ministry of the Merchant Navy, who regarded the plant as extremely harmful to the ecosystem (involving the dumping at sea of approximately 3,000 tons per day of ferrous sulfate, a processing residue). This situation fueled a bitter conflict between, on the one hand, the company and the workers who feared unemployment and, on the other, the local authorities supported by environmental committees. Between 1972 and the beginning of 1974 Montedison continued to dump ferrous sulfate into the sea, having been granted provisional permission by

Figure 4.1. The port and industrial area of Piombino.

the Livorno Harbor Authority. By 1973 the situation had become very tense: there was an attack on the ships used by Montedison for the dumping, and violent disorders erupted even in Corsica, since the ferrous sulfate reached the coasts of the French island. The Regional Council of Tuscany took on the matter, prompted by a question put to it by some Christian Democrat councillors concerned about the "economic impoverishment" of the Scarlino area if Montedison decided to close down the plant. The controversy ended on April 28, 1974, when the Court of Livorno condemned the management of the Scarlino plant for the ecological devastation caused in the Tyrrhenian Sea.

Currently the titanium dioxide plant is part of Huntsman's (a multinational chemical company), and albeit under a rigid control protocol, it continues to employ seawater in its production processes (3,651,800 m^3/year). Since 1962, Solmine (today New Solmine), also operates a company producing sulfuric acid in the Scarlino plain. Up to 1994 Solmine used mineral pyrite (later replaced by sulfur), but the processing of pyrite caused pollution from the arsenic contained in pyrite ash. Even today the presence of titanium dioxide and sulfuric acid plants makes the plain one of the most polluted areas in Tuscany.

THE DEVELOPMENT OF THE TOURISM INDUSTRY

Toward the end of the nineteenth century medical science reevaluated the healthful effects of sea air and sun. In the final glimpse of the century the spread of sand treatments (hot sand baths) opened the way to a closer relationship of man with beach. In the twentieth century those beaches began to fill up with people.[6] Until the first decade of the twentieth century, beaches lacked infrastructure and were regarded only as places to walk. The turning point came in the 1920s and 1930s when the myth of the healthy suntan made the beach a focal point for seaside life. The relationship with water also changed: saltwater bathing lost its therapeutic function and turned into a leisure activity. The discovery of sunbathing caused a revolution in the geographical map of tourism: the Mediterranean coasts (already for some decades very popular for winter tourism) began to attract a growing number of holiday makers during the hottest months. For the first time Mediterranean shores competed with the cold coasts of the North Sea, and within a changed cultural and aesthetic context, they came out the winners.

During the twentieth century the emerging Mediterranean resorts prepared to receive entirely different guests from those of the past, since elite tourism no longer represented the economic hub of this industry. The period between the two world wars not only launched the suntanning trend, but also witnessed the emergence of middle-class tourism: the average tourist no longer contemplated big hotels, but preferred less expensive solutions, such as small hotels and boardinghouses. In Italy, elite seaside tourism had been restricted to the Ligurian Riviera, Sorrento, the Amalfi coasts, and places such as the Lido of Venice and Taormina. The growing demand for middle-class tourism gave rise to the development of a second group of bathing localities, concentrated along the north-central Adriatic coast and the Tyrrhenian coast. These hosted a clientele that generally demanded less expensive accommodation, though with a degree of difference (from small boardinghouses to one or two-star hotels).

The seaside resorts of Tuscany became very popular destinations; their customers came from the upper-middle income bracket, although there were boardinghouses for the less prosperous holiday makers. Between the 1950s and 1970s Italy became the leading country in Europe for seaside tourism, attracting—in addition to Italian holiday-makers—60% of European seaside tourists. First among the Mediterranean countries, Italy was able to create a minimum level of facilities that allowed it to enter the market with highly varied accommodations and able to satisfy all income classes who were discovering tourism. The number of hotels, boardinghouses, and lodgings continued to grow decade after decade, in perfect harmony with the increase in tourists. Italy was a precursor (along with France) in seaside tourism, preceding all other Mediterranean countries and attaining a leadership which remained unchallenged until the first half of the 1980s. The other side of success's coin was the growth of coastal urban areas, characterized by expansion without regulation and resulting in an irreparable destruction of the coastline environment. In Tuscany too the coastline appears disfigured by unplanned urban development, from the illegal construction of holiday homes, to the erection of great hotels next to the beach, to the construction of yachting harbors. The land grab is a constantly growing phenomenon and is likely to further disfigure the coast (in the town of Livorno, impermeable surface increases by 0.25 km^2 each year and is equal to about 23% of the total area). Land use increases unabated even in the face of the deep transformation occurring in the tourist industry; for example, the stretch of coastline taken into consideration here is in a stage of blatant development speculation. The estate of Rimigliano, in the commune of San Vincenzo—an area (640 hectares) of great environmental value next to the coast which the municipality has not declared a protected natural area—is the scene of intense urban development with the building of hotels (about 30,500 m^2) and holiday homes (around 17,000 m^2). A new marina has also been built in San Vincenzo, which is causing rapid erosion of the beaches north and south of the harbor. Along the coastal area of the municipality of Castagneto Carducci, 60,000 m^3 of new buildings are being constructed as holiday homes, which will infringe upon the dune system of the area and on the Tombolo di Donoratico (characterized by a pine forest and a system of sand dunes) for a little less than 800,000 m^2.

THE LEGACY OF ECONOMIC DEVELOPMENT: AN AFFECTED ENVIRONMENTAL SITUATION

OVEREXPLOITATION OF WATER RESOURCES AND THE SALINIZATION OF THE WATER TABLE

The coastal area is characterized by a considerable consumption of water (generally above 150 liters per person per day): the requirement of Val di Cornia is 83,000 m^3/km^2 and that of the Huntsman Dioxide plant in Scarlino is about 7 million m^3/year. The total demand for the entire Tuscan coast is 14.7% civil, 43.5% industrial, and 41.8% agricultural.[7] From the 1970s to the present, the development of heavy industry and tourism (and, to a lesser degree, intensive agriculture) along the coast has caused a progressive increase in water consumption: water has not been obtained from surface sources like rivers because these are of a seasonal nature and their capacity varies greatly over the course of a year, with long periods of almost total dryness. Water has to be obtained by drilling wells into the deep aquifers of the coastal plains (in Val di Cornia alone, the number of wells has increased from forty-nine in 1928 to one thousand at the beginning of the 1990s). The overexploitation of water tables has caused the piezometric level to drop considerably (in some areas it fell around 10 meters from 1961 to 1984). The decreased freshwater pressure has allowed the gradual infiltration of saltwater into the water table; infiltration that started near the coastline has progressively advanced inland, affecting the water table utilized for agriculture and factory wells, which have to treat the water with lengthy and expensive desalinization.

The extraction from wells has caused an increasing imbalance between extracted water and the natural reserves of underground water. The presence of chemical, petrochemical, mechanical, and steel industries underlines the highly critical problem of water availability. It becomes even more worrisome when we think that the indicator utilized does not take into account water consumption in irrigation and for tourism, which are considerable. The increasing human presence occuring in the summer months causes excessive extraction from the aquifers, aggravating the problem of salinization and increasing chloride levels in the water up to 1,200 mg/liter (the maximum tolerated level is 200 mg/liter).

WATER POLLUTION

The area is experiencing significant groundwater, river, and seawater pollution into which effluents of industrial and residential areas spill. Sewer problems are frequently linked to water purification plants whose technical inadequacies are ineffective for the magnitude of the pollution. This contamination intensifies during the summer season when the coasts are invaded by a large number of tourists. The purification plants are geared toward the much smaller population living there year-round and cannot satisfactorily treat the waters that flow into them.

The pollution situation is made more complicated by the existence of the industrial plants at Livorno, Rosignano, Piombino, and Scarlino. The main problem is heavy metal contamination (mercury, chromium, and nickel) from industrial effluents and residues deriving from ores. Mercury pollution is caused by the ores of Monte Amiata and by geothermal activity, in particular from the geothermal power plants in Piancastagnaio and Bagnore. The other significant source of mercury contamination is the industrial plants producing caustic soda and sodium carbonate by means of mercury electrolytic cells.

The most important industrial plant is Solvay in Rosignano, which utilizes mercury catodic cells for the production of chlorine, hydrogen, and caustic soda: the infiltration of mercury into the sea was huge until 1976, but some still occurs today. Water contamination also derives from older mercury pollution residue settled beneath the deeper marine sediments. In addition, the Solvay plant discharges significant amounts of calcium carbonate and sodium chloride into the sea (the waste flow rate was estimated at about 200,000 tons/

year in 2003), which are utilized in the production of the Solvay soda. The result is a "white spot," a portion of the sea and of the beach in front of the chemical plant which has taken on a Caribbean-type color (the beach is white and the water is a transparent blue) and appears as a white or bright spot along the coast. Sources for the chromium contamination are mining, the production of metal alloys, and of finished products containing chromium, such as dyes, paints, and chrome-plated objects. The main sources of anthropogenic nickel contamination are the mining of the metal, the production of nickel alloys, and its use in finished products.

COASTAL EROSION

The massive pumping of groundwater from aquifers intensifies the effects of subsidence, which is happening in many coastal areas and represents a risk for the land adjacent to a coastline that lies almost at sea level, at times even lower. In addition to intensifying subsidence, human activity has contributed to erosion along the coast. The building of piers, breakwaters, and other structures has changed the processes of sediment deposition and is causing serious erosion. Many beaches were historically replenished by sediments from river estuaries rich in lime. Reclamation processes in many swampy areas and the excavation of streams—with the intent to prevent flooding and for utilizing gravel and sand for construction—have reduced the accumulation of sediments, causing erosion and a retreat of the coastline, which in parts of the Gulf of Follonica extends up to 200 m. Such retreat has also reached worrying magnitudes in some stretches of the coast around Vada Marittima and Marina di Cecina.

The human impact on such erosion is even larger in tourist areas. In the summer months dunes become parking lots for cars and campers. The underbrush is devastated or destroyed altogether, thereby exposing the soil to rapid erosion. In many areas of the coast, consolidation efforts based largely on artificial beach nourishment are practiced, but the severity of the problem and possible future scenarios (further retreat of the shoreline and the flooding of reclaimed land) expose the necessity of adopting planning policies for the region which must carefully envisage an environmentally responsible use of the coastland, including the exclusion of certain areas to development, the protection of dunes, and the prohibition of quarrying riverbeds for inert materials.

MORE ENVIRONMENTAL PROBLEMS

In addition to these issues, there are other pressures on the environment which cause a more or less marked deterioration of the ecosystem. The industrial areas of Livorno, Rosignano, and Piombino are affected by widespread air pollu-

Figure 4.2. The white beaches in Vada. In the background is the Solvay chemical plant in Rosignano.

tion caused by industrial plant emissions (sulfur dioxide, nitrogen oxides, and particulate matter).[8] There is also the problem of industrial waste (lead and granular ashes, wastewater treatment plants' sludges, and water decarbonation sludges). The question of waste disposal is especially critical in the plain of Scarlino where the Huntsman Dioxide plant produces significant amounts of red gypsum derived from the desulfurization of liquid and gaseous effluents from the production of titanium dioxide. Red gypsum is disposed chiefly in the area around the plant and in a dump in the region of Follonica.

The continuity along the whole coastline between industrial and tourist-residential urban areas means that the pressures exerted by industrial sites frequently affect tourist bathing areas (e.g., the blatant effects of the Solvay plant, whose activity has repercussions on a coastline area where beaches are called *bianche* [white] because of the plant's sewage spilling into the water; the fallout of dusts—so-called *spolverino*—from the steelworks of Piombino).

The combination of industrial activity and tourism also creates detrimental effects on natural resources, compounding the effects which one or the other produces alone. The most emblematic case is the joint pressure on groundwater from coastline industrial plants and agriculture, combined with water pumping for municipal needs. In the summer, when tourism is at its highest, the water crisis peaks as the various factors act synergetically.

* * *

The economic development model characterized by the combination of heavy industry and intense tourism on the coast has generated a high density of businesses, residential areas, infrastructure, and overall, an overbearing human load. The development of modern industry in Livorno, Piombino, Rosignano, and Scarlino first drew workers from the surrounding areas and then enticed a lot of people from inland areas to the coast. The lure of industry has joined that of beach tourism, providing employment and good economic growth prospects, to heighten migration to the coast.

Human activity has resulted in overconsumption of natural resources, pollution of the water and air, occupation of previously undeveloped land, and other forms of pressure which have often pushed the coastal area beyond its carrying capacity.

The decline of industry began in the 1980s, and the resulting loss of jobs forced the diversion of the workforce toward other sectors. Seaside tourism along the coast—too concentrated in time and space—is not only incapable of providing those new jobs, but has generated such congestion that it requires an intervention to decrease its anthropogenic effects in the summer months. The coastal areas demand further interventions to restore the quality of the environment by reducing erosion, restricting pressure on aquifers, recycling water, and restoring its quality.

PART II

MOBILITY AND THE ENVIRONMENT

Historiography in Italy has paid (and still pays) little attention to topics con-cerning the interactions between mobility and environmental problems.[1] Historians have begun only in recent years to address in detail the environmental problems caused by transportation. Until now their attention was directed mainly to air pollution; since the 1960s, the United States, Western Europe, and Japan have addressed the environmental impact of transportation (in particular motor vehicles) by passing legislation to reduce air pollution. Other issues, however, such as the depletion of natural resources (particularly water), have been neglected. There are no studies, for example, on the environmental impacts (on the city and surrounding areas) generated by the flow of goods in and out of factories. In recent years we have seen an attempt at creating a new interdisciplinary sector, environmental mobility studies, to bring together all types of transportation, environmental, and social history, studying the many factors of mobility from an environmental point of view.[2] A first attempt in Italy was made by a special issue of the journal *I Frutti di Demetra* devoted to "The Car and the Environment."[3]

In Italy the car has represented not only a means of mobility and a technological product that revolutionized transportation on land, but it has been a player in social transformation, a symbol at once of freedom, independence, well-being, and progress. In the aftermath of the Second World War the car even contributed to that process of national unification which the railroad had begun more than a century before. While television broke down language barriers, the car not only shortened physical distances, but created a new symbology that became rooted in the collective imagination; Fiat runabouts were the most visible symbol of that desire for modernization which had guided the transformation of the Italian economy and society in the postwar period.[4]

In Italy the automobile began to be seen as a driving force, so to speak, toward environmental change only at the end of the 1970s. This was no great surprise since Italy had been a latecomer: in 1950 there was one car for every 81.9 residents, compared to 48.7 in West Germany, 17 in France, 15.2 in Great Britain, 5.8 in Australia, and 3.1 in the United States. In the 1950s and 1960s the media and public opinion sporadically looked at the problem of landscape disfigurement from to highways and traffic noise, but not yet at the atmospheric pollution caused by internal combustion engines. There was a widespread opinion that environmentalism was merely an attempt to preserve a privileged lifestyle and certain cultural values enjoyed by a minority elite, in other words, a luxury that only the comfortably well-off could afford. This is why environmental policies were considered far less urgent than those aimed at ensuring a better standard of living and a minimum social security for those great segments of the population who were still living in destitute conditions. The environmental impacts produced by cars and other motor vehicles were also ignored by the manufacturers, who tended to minimize the concerns raised by scientists. Cars were turning out to be a highly profitable business, with an enormous market penetration, and the industry was far from willing to see its earnings drop by investing money to equip vehicles with antipollution devices that did nothing to stimulate sales.

The first change was precipitated by the international oil crisis of 1973. That year in Stresa the Conference on Traffic and Circulation (the foremost Italian consultative body on motor vehicles) approved a resolution calling for a greater commitment to fighting pollution.[5] The motor vehicle press—whose attitude was constantly seesawing between antipollution measures and defense of manufacturers' interests—began to give increasing space to the problem of pollution caused by motor vehicles. In early 1977 the monthly magazine *Quattroruote* published two articles ("The Truth about Hydrogen Engines" and "Hydrogen Is Not an Opinion") in which it was pointed out that the first hydrogen engine patent, filed by the American A. F. Burstall, dated back to 1927 and that an Italian engineer (Massimiliano Longo) was achieving good test results with a hydrogen combustion engine on an Alfa Romeo Giulia 1300. In December 1977 the magazine reported on the electric motor ("The Electric Vehicle: How and When?") on which Fiat, Ansaldo, Enel, Garelli, Magneti Marelli, and Piaggio Zagato were working.

The watershed moment came in 1986 when the Italian Ministry for the Environment was established and the Single Europe Act formally legitimized community environmental policies—already begun with the environmental action programs of 1973–1977, 1977–1981, and 1982–1986—by introducing a new section (VII) into the Community Economic Treaty expressly dedicated to the environment.[6] The European Community then began to play a fundamentally important role in developing environmental legislation in its member states, including Italy, constantly stepping in as an intermediary between the national stage, very often too limited, and the global, which lacked the institutional possibilities for binding action.

From the 1990s on the environmental problems caused by motor vehicles (especially those concerning atmospheric pollution and the uses of land) were also regulated by European Union directives, whose importance is particularly evident in the case of Italy: in the absence of European restrictions, the resistance to an effective body of environmental legislation would have been far greater. Without the direction of the European Union with which Italy had to conform in accordance with signed treaties—the pressure groups that had delayed application or voided the content of environmental laws promulgated between the second half of the 1960s and the early 1980s would have continued to procrastinate over approval of incisive regulations.

The essays in this second part (chapters 5–7) tackle certain problems that arose from motor vehicles. Chapter 5 deals with the transformations of urban space and contains three brief case studies in Rome, Milan, and Florence. Environmental modifications brought about by the car posed wholly new problems. With the advance of the "motor revolution," the physical space of Italy altered radically; the landscape was greatly modified by newly built highways and other road infrastructure. Urban space itself, furthermore, was redesigned to function with the new mobility: viaducts, road intersections, parking lots, service stations and garages, supermarkets designed to be "car friendly," streets and piazzas cluttered with parked cars, the new "tenements" in the suburbs with parking spaces in front of them. All of these defined the architecture of towns in the ongoing expansion. Chapter 6 deals with environmental problems linked to the increase of road transportation in the second half of the twentieth century, and inserts into the historical narrative data on materials flows, virtual water, average amount of waste per worker, and the emission of polluting substances. Chapter 7 is dedicated to policies addressing the control of atmospheric pollution, which represents the fulcrum of the debate on interactions between motor vehicles (cars in particular) and environmental transformations.

CHAPTER 5

THE ENVIRONMENTAL IMPACT OF THE CAR ON ITALIAN CITIES

1950–1974

The earliest signs of the remarkable urban growth which pervaded Italian cities, particularly in the regions of the center-north, during the second half of the twentieth century became visible in the early 1950s. Mass migration occurred from the southern regions to the cities, where the leading productive sectors were causing rapid economic growth. It was deemed the *miracolo economico* (economic miracle), which reached its peak between 1958 and 1964. Not only the great industrial centers of the northwest attracted immigrants, but also the cities of the northeastern, central, and parts of industrialized southern Italy.

URBAN GROWTH AND BUILDING SPECULATION

Between 1951 and 1971 the twenty-three metropolitan areas of the country[1] experienced a population growth of almost 12 million in 1971, while occupying only 8.3% of the national territory; they also housed 49.4% of Italy's entire population. The metropolitan areas of the center-north doubled their populations (from 8,418,296 in 1951 to 16,766,709 in 1971), whereas in the center-south growth was less dramatic (from 6,345,205 to 9,934,832).[2]

Under pressure from this unprecedented migration, reconstruction of the cities came under the control of private speculators and "builders clubs," which catered to the interests of builders, construction industry professionals, real estate owners, private speculators, and investment banks.[3] It has been pointed out that such clubs extended their interests to all economic and social sectors, in line with the well-known concept that "if the building industry thrives, all economy thrives."[4] It has also been calculated that between 1956 and 1965 investment in the building sector represented 12.5% of Italy's GDP and that housing prices rose 250%, whereas the general increase in consumer prices was 140%.[5] During the reconstruction years and the period of miracolo economico, the construction industry, together with the car industry, became the chief sector in the Italian economy. Clearly this had been the result of a precise political calculation designed to enable the construction industry to develop unimpeded, since it created jobs for a considerable number of otherwise unemployed workers (the building industry employed more than 2 million workers) while simultaneously providing accommodations for the homeless.

During the 1950s and 1960s urban growth was heavily influenced by private interests, which at the same time delayed and conditioned urban planning. Cities thus ended up haphazardly spreading out while building speculation produced, over the years, a widespread lack of public amenities; apart from depriving new developments of the recommended "settlement fitness" specifications set by planners, that also created densely built urban areas which lacked essential services.

The pace of growth was chaotic. By the end of the 1950s there were more than 47 million rooms (for a population of 50 million), the majority of which were built in regions of the center-north where there was practically one room per resident. In the 1960s, during which 16 million rooms were built, the expansion of housing was even more conspicuous, especially in medium-size towns. The number of dwellings, including those built illegally, grew inexorably throughout the 1970s. During this decade, compared to a population growth of 2 million, 23 million rooms were built, and owner-residents reached 60% of the total.[6] What happened in Italian cities between 1946 and the end of the 1970s is very effectively summed up by the planner Giuseppe Campos Venuti:

> The cities of the miracle growth, by virtue of a phenomenon called by architects "the oil stain," meaning spreading in all directions, indiscriminately invaded the countryside, not leaving aside indispensable space for future gardens or services: one mighty block after another, interspersed by narrow gaps where streets are but narrow lanes, and with a density which only fulfills the arrogant wishes of building speculators. The communes watch powerlessly: while proprietors "allot" and sell without leaving aside one single square

> meter for the amenities of collective interests, most of the time they demand to be paid for land where the communes will later have to build roads, drains, electrical networks, water, and gas pipelines.[7]

Within this context, the proliferation of private transportation was regarded as indispensable in order not to hamper building expansion. The car, in fact, enabled an ever-growing number of people to reach the new suburbs in comfort and at a reasonable cost.

THE CAR "BOOM"

The growth of private transportation went hand-in-hand with the building expansion. In Italy the car had not had a significant influence on the organization of public transportation, or upon personal mobility, until the early 1920s when the Fascist regime started the first highways, effectively boosting private transportation.

It was only after the Second World War that the explosion of the car gained new fuel: in 1946 private cars numbered 149,649, a number that increased to 342,021 in 1950. In 1950 production, which in 1945 had dropped to 2,093 units from the 1939 level of 55,553, reached 101,310 cars. Notwithstanding the first symptoms of an economic recovery, the density rate of cars in circulation, that is, the number of people per vehicle, continued to be very low. In 1950 there was one vehicle every 81.9 residents, compared to a rate of 48.7 for West Germany, 24.2 for Switzerland, 20.7 for Belgium, 17 for France, 15.2 for Great Britain, 5.8 for Australia, and 3.1 for the United States.[8]

In 1959, despite the increase in cars, at an annual rate of 15.48%, which had taken place from the start of the decade, there was still no mass motorization in Italy. It continued to be essentially a country of "two wheelers" or "three wheelers," those typical little transit vans that clogged urban traffic. Until the economic miracle years the expansion of private vehicles was characterized by an extraordinary increase in the sales and circulation of scooters.

Italy joined the club of motorized countries only in 1964, at the end of an extraordinary period of growth already known as the miracolo economico. In this year the number of cars surpassed that of scooters: the four-wheeler (4,674,644) overtook the two-wheeler (4,656,035) by a mere 18,609 units. In the economic boom years, between 1959 and 1963, car ownership grew alarmingly, with a mean annual increase of 20.7%. Car density went from 81.9 to 9.9 residents per vehicle in 1964, and the number of cars on the road had more than tripled from 1958. From 1964 to 1974 there were 14,303,761 vehicles on the road, and the mean density had reached 3.6 residents per all types of vehicle, 3.89 per car. Therefore 1974 marked the peak of the first phase of mass dissemination of the private car. Beginning in 1975 commercial trends altered considerably, and the production of such car models as had hitherto characterized the country were all but taken out of the market, replaced by new designs to compete with aggressive foreign competition. By the end of what may be rightly called the "car revolution," Italy had one of the highest density of vehicles in the world, having reduced to nothing its initial gap. The exceptionality of the Italian case was confirmed by continued expansion after 1974. In 1990, for example, Italy had the highest density of cars in Europe and was fifth in the world following the United States, Canada, Australia, and New Zealand.

CITIES "UP TO THEIR NECKS IN CARS"

Following the sudden explosion of car sales between the end of the 1950s and the beginning of the following decade, traffic became problematic in the principal cities of Italy and in many of its towns. In 1968 an important weekly magazine, *Domenica del Corriere*, gave ample space to the chaos which pervaded the roads of all Italian cities with an article titled "Italy Up to Their Necks in Cars."[9]

During the first half of the 1960s this situation was caused by local administrations not able to make the right choices for urban planning following growing demands for mobility; instead they restricted themselves to vaguely pointing out the problem, for essentially two reasons.[10] The first was the method used in their decision making. Statistical forecasting techniques aimed at predicting traffic density growth, were practically unknown in Italy, or at any rate they were rather obsolete since they did not consider political and administrative issues in urban planning (although these matters were regarded as priorities by researchers from English-speaking countries). No organized traffic studies were available; the scanty data were restricted to analyses that were mere collections of quantitative data concerning vehicles in circulation. The few inquiries on the origin and destination of vehicles in transit, carried out by the Automobile Club Italia (ACI),[11] were of little value since they did not take into account the reasons for movement and the use of road space.

The second reason concerns the absence of adequate legislation on the matter. In fact, the rules concerning land transportation networks were still the ones addressed by the 1942 planning law[12] (passed at a time when in all of Italy there were only 73,790 cars). Parliament had never been able to pass a new general planning law. As a result there was a serious lack of space destined for public amenities (especially road networks and parking spaces), urban structures progressively deteriorated, and transportation system were congested.

During the first half of the 1970s, the congestion of Italian cities remained substantially unchanged; measures adopted by the municipal administrations, including the interventions of traffic engineering and adaptation of the infrastructures (the only ones that had found the unanimous consent of the administrators), failed. To solve the urban traffic problems, the Associazione Nazionale Comuni Italiani (ANCI; National Association of Italian Communes)[13] proposed in the pages of its news magazine, "the possibility of coordinating infrastructure systems in large metropolitan areas with territorial and regional ones . . . , the development of systems of urban public transportation based upon fast vehicles . . . the study of efficient underground network systems, the planning of a network of urban toll and free highways, the realization of large parking lots near city centers."[14]

The remedies put forward by the Lega Nazionale dei Comuni Democratici, Regioni, Province ed Enti Minori (National League of Democratic Communes, Regions, Provinces, and Minor Institutions)[15] were not very different from those proposed by ANCI. This organization of socialist, communist, and republican communes proposed the rationalization of taxi ranks and bus stations, the construction of underground passages and parking garages, the building of intercity bus stations, the ruling that new roads should not be less than 14 m wide, the institution of municipal offices for monitoring traffic, and the organization of traffic on the basis of speed, type of vehicle, and weight.[16]

Measures for regulating traffic actually enforced by local authorities chiefly concerned traffic flows (one-way streets, no left turning, and the installation of automatic traffic lights) and parking (the establishment of "green zones," regulating the timing of parking, pedestrian precincts, and the installation of parking meters). Some local administrations had instituted preference lanes reserved for public transportation, with the purpose of easing traffic circulation and to increase the speed of commercial traffic (hampered by cars, buses were restricted to an average speed of 15 km/hr). Even so, the lack of methodical planning and the scarcity of reserved lanes resulted in dashed expectations. The failure of traffic engineering measures was essentially due to their fragmentary and casual nature. Local councils adopted these measures not according to a careful plan, but as a result of pressures to intervene, which often resulted in emotional reactions from the indignant man on the street who was fed with the urban traffic situation. The measures adopted by local administrations also did not follow commonly shared criteria, or even scientific criteria. This chaotic situation led experts to demand the passing of national laws dictating traffic planning guidelines to which local authorities must adhere. Lacking such legislation, many communes, chiefly those of small and medium size, adopted no measures at all to deal with the traffic chaos; in the cities such measures were applied rather irrationally and casually.

Large cities, beyond planning traffic engineering, applied other measures of a structural nature. Generally speaking these consisted of bypasses and ring roads, whose final aim was to lighten traffic in historical city centers. In a few years, however, high-density developments and businesses were built along these new roads, turning bypasses into urban avenues with traffic congestion problems similar to or worse than the original ones. Furthermore, often such bypasses and ring roads were planned with inadequate technical features for supporting the density of their traffic at that time and therefore were absolutely insufficient for absorbing the ever-growing number of cars in circulation. The budget allocated to road building in Italy went almost entirely to highways, whereas urban and ordinary roads received only negligible resources, half of what France and Britain budgeted for such roads and one-fifth of Germany.

Restricted to larger cities such as Milan, Turin, Genoa, Venice, and Rome, there were ambitious plans for urban underground or aboveground parking garages and metropolitan railways (among these only Milan and Rome got their underground rail lines). During the course of the 1970s, town councils continued to postpone any policy applying precise and rigorous rules to urban traffic. Policies remained restricted to limited measures to make the best out of what already existed and to rationalize traffic, as far as possible, in an attempt to prevent the total breakdown of urban circulation.

URBAN CHAOS, CIRCULATION CRISIS, AND POLLUTION: THE CASES OF ROME, MILAN, AND FLORENCE

Rome, Milan, and Florence represent cases in point for how local administrations confronted the problem of urban expansion and increasing traffic. Planners speak of a true "town-planning massacre" which took place in Rome, chiefly from the second postwar period to the 1960s.[17] During this period the city was subject to a savage building speculation rush, with the consent of a center-right administration and by the vested interests of the aristocracy, a builders' bourgeoisie, and the Vatican-owned Società Generale Immobiliare.[18] A comprehensive operational plan was not developed by the city council until 1962 and was approved only in 1965, when the city had already been overbuilt with disregard for any principle of fair administration.

This speculation exercise was accompanied by massive widespread illegal building, fueled by the demands of a fast-growing population (an influx of 623,100 more residents between 1955 and 1970). In 1962 illegal buildings occupied 3,800 hectares; in 1977–1978, during attempts to alter the general plan of 1962, illegal building amounted to 267,000 rooms spread over 5,000 hectares. The dramatic lack of essential amenities, the scarcity of greenspaces (2 m^2 per resident), and traffic which was impossible to govern, created a situation apparent to all. Exacerbating this situation was the slow progress in building the subway, the dismantling of tram lines (from 182 km in 1953 to 43 km in 1966, with a decrease from 396,320 to 77,897 passengers), and the inadequacy of the road system, which was still practically that of the ancient Roman roads.

An article appeared in *Quattroruote*, Italy's leading monthly car magazine, which very effectively described the lack of planning and spoke of an "almost comical example" of the enforcement of regulations established by the general operational town development plan:

> An example of this . . . is the "bottle-neck of the Via Salaria." In order to ease the heavy traffic of cars and trucks on this road, the town-development plan contemplated a one-way traffic flow to be made possible by the building of an alternative route running behind the mansion of an aristocrat. As it happens, on this site a builders' cooperative had erected two luxury villas, with the result that now the authorities of the Capitol seek the key to the solution of a problem which has become insoluble, by means of the most audacious experiments.[19]

In 1966, when the Province of Rome had 624,009 cars in circulation (increased to 1,191,547 in 1974), the average speed of urban traffic was 5 km/hr.[20] In spite of the institution in 1962 of a Citizens' Consultative Committee on Traffic (Comitato Cittadino Consultivo del Traffico), the city administration, under the pressure of public opinion, renounced the adoption of a serious policy to redress the imbalance of the ever-more congested transportation system in favor of public transportation. The proposed remedies put forward by the Commission of Inquiry on Traffic (Commissione di Indagine sul Traffico), established in November 1963, remained words on paper. The report spoke of "green-wave routes" (*itinerari ad onda verde*), some kind of "urban highways," of a traffic plan aimed at establishing priority for public transportation and rigid parking regulations, of a "specialization" of traffic areas including "pedestrian precincts."[21] The blatant inconsistency of the traffic policy put forward by the Capitoline administration—which, written in a report by the commission, should not have "altered the existing urban equilibrium" or oppose "in any way" the "development of the city as envisaged by the operative plan"—was that it failed to create a pedestrian-only area between Piazza del Popolo, Via della Mercede, Via del Corso, and Trinità dei Monti. The plan was aborted after only a ten day trial period.[22] In 1968 the administration, notwithstanding the violent reactions of the public (43% of all journeys were made in private cars, 40% on public transportation), decided to close a portion of the historical center to traffic and in 1973 the Villa Borghese Park. Such restrictions remained largely ignored by drivers.

Rome also held the unenviable record of being the city with the highest level of pollution deriving from motor vehicles. Already in 1960, both in the Via del Corso and in the underground passage by the Quirinale, a concentration of carbon dioxide of 0.3–0.5 parts per thousand was measured, and it was calculated that the level would stabilize around 1.5–2.0 parts per thousand.[23] In 1967 the vehicles in circulation, most of them obsolete, released into the air 269,040 tons of carbon dioxide, 31,289 of hydrocarbons, 11,700 of nitrogen oxides, 873 of sulfur dioxide, 383 of organic acids, and 1,074 of particulate matter.[24]

In Milan the first operational general town planning program was approved in 1953, when large sectors of the city already had a very high density of residential units, which was favored, as in Rome, by the reconstruction plan.[25] The 1953 plan displayed all the negative traits of the town planning methods in those days: a high residential density—the plan envisaged 2.5 million rooms for a population of 1.28 million, though it did not actually reach beyond 1.75 million in 1973—overblown growth forecasts, which represented a concession to private speculators; the phasing out of industrial developments and the resulting conversion of central areas for services; and the lack of public amenities, in particular greenspace (1 m^2 per resident in 1961).[26] Under pressure of speculators, addenda to the general plans in 1963 and 1969 failed to give results, and the situation remained unchanged until the publication of a new plan in 1976. Irrational traffic planning led to the development of a dense network of roads, a chaotic system of circulation, and inadequate parking. Furthermore, in Milan too the enthusiasm for the private car led to the dismantling of tram lines, which fell from 302 in 1953 to 128 in 1966. In a belated attempt to find a remedy to this situation, the administration of Milan was one of the first to adopt measures to ease traffic congestion by forbidding parking between certain hours of the day: from 7:30 to 9:30 a.m. and 1:30 to 3:30 p.m. (the so-called "green zones"), and by allocating certain areas of the city center for pedestrians only. Notwithstanding such measures and the building of a metropolitan subway system, traffic problems remained practically unsolved. In 1964 in Milan Province there were 469,552 cars, which in 1974 had increased to 1,233,285 (8.63% of all cars in circulation in Italy). Every day in 1972, 342,000 cars drove into the city from suburban districts, bringing in 74% of all commuters, whereas 670,000 cars moved daily from the hinterland municipalities to the city center (only 33% of commuters used public transportation). Milan was, after Rome, the city with the highest level of air pollution emitted by cars. In 1955 it was shown that the air of central streets contained 0.48–2.24 mg of lead per g of particulate matter, and the air of peripheral streets had 0.40–0.88 mg.[27] In 1967 vehicles discharged into the air 164,160 tons of carbon dioxide,

Figure 5.1. Roman traffic jam in the 1960s.

31,295 of hydrocarbons, 6,815 of nitrogen oxides, 533 of sulfur dioxide, 234 of organic acids, and 655 of particulate matter.[28]

The city of Lombardy in 1972 had a level of sulfur dioxide of 0.0196 kg/m^3, compared to a national average of 0.0019 kg/m.^{3} Due to Milan's poor environmental situation, in 1969 the environmental organizations Italia Nostra, the Centro Studi Attività Politiche (Center for the Study of Political Activities), and Ente Provinciale per il Turismo (tourist office), organized the first important public awareness campaign on air pollution, called Aria per Milano (Air for Milan).

Until the mid-1950s Florence, not yet having an operational development plan, expanded in a disorderly fashion.[29] From 1958 its urban layout was further undermined by the approval of an operational plan allowing intensive development in the plain (with a density of $7m^3/m^2$ to $20m^3/m^2$) and building on the hills, leaving the city center without real protection. Furthermore, planners allowed only small areas for public greenspace and services. An attempt to correct this situation was made by the Detti Plan of 1962—named after the councillor responsible for urban planning, a member of a center-left coalition, who promoted it—which established as a priority the protection of the city center, reduced building density, increased the area designated for public greenspace and services, and directed the expansion of built-up areas toward the northwest. After the crisis which, in 1964, led to the fall of the center-left coalition in the local government, the Detti Plan was de facto shelved without raising any public protest. The numerous addenda to the plan, approved in the 1960s and 1970s, encouraged substantial development, in spite of a relatively modest growth in population, which led to the doubling in size of the urban area by adding 220,000 rooms (a 45% growth). In this situation the road network too, along with public greenspace and service areas, suffered—to the benefit of building speculation. The decided shift of the city toward services accompanied by an inadequate transportation system caused traffic congestion problems well in advance of other medium-size cities. Between 1952 and 1964 in the Province of Florence private car ownership rates increased by 634.33%. In 1964 there were 163,308 vehicles on the road (147,614 of which were cars), one car for every seven residents. In 1974 it increased to 428,135 vehicles (397,066 of which were cars), one car for every 2.77 residents.

One should not, therefore, be surprised that aerial photographs taken in the 1960s show the roads of Florence occupied by a great snake of cars and that squares were only huge parking lots, with cars parked right up to the steps of the churches of Santa Croce and the Cathedral of Santa Maria del Fiore. In 1964 borough commissioner Lorenzo Salazar tried to rid the city center of cars with a bill banning them from the medieval heart of the city. A protest led by the Christian Democrat MP Salvatore Foderaro, president of the parliamentary group Amici dell'Automobile (Car Friends), induced Salazar to withdraw his bill.[30] In 1971, despite the protests of traders and residents, a limited traffic zone, called *zona blu* (blue zone) was enforced. Over the years, in spite of work on existing structures and new roads, the traffic situation did not improve. Various trunk roads appeared disjointed, and car circulation remained restricted to the system of avenues which crossed the city from its southern outskirts (Rovezzano) to the north (Peretola). In place of a true ring road, road belts were built connecting the highway to the city. Both the lack of an appropriate system of parking lots and the dismantling of tram lines (from 65 km in 1953 to none in 1966) aggravated the existing traffic problems. Florence appeared to be one of the most polluted cities in Italy. A survey carried out by the Istituto di Chimica Analitica of the University of Florence found that road traffic and domestic central heating systems were the chief sources of pollution.[31] Measurements taken in fall 1972 and winter 1973 showed a high percentages of sulfur dioxide (0.0035 kg/m^3) and the presence of calcium, aluminum, silica, magnesium, manganese, iron, vanadium, copper, zinc, and nickel, in addition to high levels of lead (an average of 7μ/m^3) and polycyclic aromatic hydrocarbons from vehicle exhaust.

Figure 5.2. Piazza della Signoria, Florence, in the second half of the 1960s.

The situation illustrated by these three cities applied not only to all of Italy's medium-size and large cities, but often to centers of more modest size; so much so that in 1973 the first report on the country's urban environmental situation of the country produced an alarming picture.[32]

THE DISCOVERY OF POLLUTION

Italy had officially discovered pollution in March 1971 when the president of the Senate, Amintore Fanfani, promoted the institution of a Comitato d'Orientamento sui Problemi dell'Ecologia (Orientation Committee for Ecology Problems).[33] In 1973 the conclusions of the *Prima relazione sulla situazione ambientale del paese* (First report on the environmental situation of the country) provided the first scientifically relevant data concerning the "state of health" of urban areas.

According to the document, Italian cities continued to occupy the last position within Europe on the relationship between the number of residents and greenspaces: 2 m^2 per person in Rome, 1.5 in Turin, 1 in Milan, and 0.58 in Naples (where in 1973 twenty-seven residential districts did not have any greenspaces at all). In Florence the inhabitants of new residential districts, when they had any at all, had only about 0.70 m^2 of public greenspace per person.[34]

The noise pollution in various Italian cities was decidedly higher than the standards regarded as acceptable (up to 50 dB) and appeared to be produced for the most part by traffic. In the streets of Milan values of 85–90 dB had been recorded in 1959; Rome reached levels of 60–100 dB in 1968–1969; and in Bologna it was 85–100 dB in 1972.

Beyond industrial fumes, the chief sources of air pollution were domestic central heating (between 1966 and 1971 the consumption of fuels had practically doubled, going from 114,109 billion kcal to 225,934 billion kcal) and motor vehicles. It was estimated that in 1968, 42% of the population lived in "high-risk" areas,[35] and that direct and indirect deaths from diseases linked to air pollution amounted to 66,799 (with 10,371 deaths due to direct causation). Cases of disease caused by air pollution amounted to 6,500,500; hospital cases and hospital days were, respectively, 80,000 and 13 million; and 3.6 million working days were lost. Between 1966 and 1971 motor vehicle emissions had increased by 46.3%, particulate matter by 6.7%, sulfur dioxide by 6.5%, nitrogen oxides by 7.2%, hydrocarbons by 7.6%, and carbon dioxide and lead by 8% each. In 1972 motor vehicles produced 5.425 million tons of carbon dioxide, 746,000 of hydrocarbons, 215,000 of nitrogen oxides, and 20,500 of particulate matter. In almost all urban areas with heavy traffic, the level of nitrogen oxide reached 100–150 ppm, while the allowed limit was 20–50 ppm. The blood of people "not exposed by occupational causes to lead-rich environments" still contained high concentrations of lead, 20–80 μg per day in excess of the limit, set at 8 μg/100 ml of blood. There were no available data on nitrogen oxides and oxidizing substances.[36]

In the 1950s the car was singled out as one of the main causes of air pollution. The public's attention had sporadically been drawn to the problem of traffic noise, but not yet to the air pollution caused by motor vehicles. There was a widespread opinion that concern for the environment was merely an attempt to preserve the privileged lifestyle and cultural values of the elite. Protection of the environment was regarded as a luxury affordable only by the few who enjoyed a high standard of living. Therefore environmental policies were thought of as being much less urgent than those concerning the improvement of living standards and a minimum social welfare for the larger sectors of the population who still lived in poverty.

In November 1957 the WHO held a Congresso per l'Inquinamento Atmosferico (Convention on Air Pollution) in Milan, in which fifty delegates from twenty-three European countries took part. From this congress came data which *Pirelli* magazine described as being "more alarming than anyone could envisage." For the first time cars were identified as "a source of danger," and the studies presented at the conference showed that in certain areas of Milan the levels of carbon dioxide in the air were similar to those of Los Angeles, one part per 4,200.[37] Beginning in 1960 the subject began to be discussed in the specialized press, which examined the relationship between the inhalation of car exhaust and certain diseases. The journal of the Commissione per l'Automobilismo Industriale (Commission for Industrial Vehicle Transportation) of ACI, in its first issue of 1961, underlined the role of motor vehicles as sources of air pollution and listed the damaging effects of smog: "To the sources of pollution, which have become increasingly more concerning over the past decades, another has been added recently: not less than one third [of pollution] is caused by the emission of products issued from internal combustion engines. . . . The effects of smog and of fog may be: 1) truly toxic and even cause death, 2) irritating to the respiratory system and the eyes; 3) disgusting or disturbing to the organism."[38]

The following year, the ACI weekly magazine pointed out that in Italy there was no "law against gases," whereas the monthly magazine of the Istituto Sperimentale della Commissione Strade (Experimental Institute of the Commission for Roads) of the Italian Touring Club published an article with concerned undertones listing diseases linked to air pollution.[39] In the September-October 1964, issue of *Annali della sanità pubblica* (Public health annals), the minister of health published a study by Professor Cominelli which unequivocally accused traffic as one of the principal causes of pollution: "Motor traffic constitutes, as we know, one of the chief sources of urban air pollution. We have tried to establish statistics concerning global pollution, translating these into values which range from modest . . . up to 25% in larger cities. . . . Whatever the volume of pollution deriving from traffic, we cannot avoid attributing to motor vehicle exhaust a particularly dangerous quality."[40]

The expert at the Ministry for Health believed "ill conceived" any optimistic attitude in this respect, and although believing it to be a "vaguely fictional" hypothesis, he regarded the "generalized use of different sources of energy from the present"[41] as the only solution which would radically deal with the problem. Notwithstanding the ongoing debate in the country, the first air pollution law was passed only during the second half of the 1960s.[42] Motor vehicle legislation consisted of mere generic statements. The first ruled that motor vehicles must not "issue polluting discharges, however caused"; the second ruled that diesel engine vehicles must not "issue fumes of a density above those established by regulations." The only concrete measure adopted by the text was that regular testing of motor vehicles should include polluting discharges. On the installation of devices to reduce toxic exhaust gases, the regulation established that the Ministry of Health should make this obligatory, but only after having asked for many opinions.

In 1967 the monthly magazine *Strade e traffico* (Roads and traffic) published a soundly based article on the relationships between air pollution and the development of urban traffic (quoting a study carried out by the Chase Manhattan Bank), highlighting the fact that transportation constituted the chief source of pollution (59.8%), followed by industry (18.7%), power plants (12.6%), domestic central heating plants (6.2%), and waste incinerators (2.6%). Furthermore, motor vehicles represented the chief source of carbon monoxide emissions.[43] The magazine highlighted the fact that in Italy the problem inherent in gasoline engines had not been studied from a "really practical point of view," and pledged that European countries should adopt rigorous standards, bearing in mind the urban planning situation, traffic, and the development of motorized transportation, in order to "grant absolute safety for public health." In March 1969 Italia Nostra sponsored a meeting in Milan on pollution caused by cars, an issue which resonated throughout Italy.[44]

Notwithstanding the vibrant tones of the internal debate and the growing alarm about the unstoppable degradation of the urban environment caused by the astonishing increase in cities' traffic and air pollution, only in 1971, following a European law (No. 70/220/CEE), did the Colombo government pass a bill checking exhaust pollution caused by motor vehicles using gasoline.[45] The bill added to the requirements to license a vehicle the limiting polluting exhaust and established that such fitness should be tested in the presence of an official from the Ministry of Health or from the Higher Institute on Health. Though the law contained general rules restricted to the acceptance of European guidelines, it failed to tailor them to the Italian situation; this made the provision little more than a declaration of intent.

Despite the progressive deterioration of the situation and the hopes of experts, during the 1970s and 1980s adequate legislation on the matter was still lacking, no national survey project had been conceived, and no urgent measure was adopted to lower the levels of toxic substances present in vehicle exhaust. There were two chief reasons for such an underestimation of the environmental impact of the motor vehicle: On the one hand the builders were so powerful—in spite of the overwhelming evidence of the relationship between car exhaust and increasing urban air pollution—they were able to postpone the enforcement of all recommendations put forward by the experts. On the other hand, municipal administrations gave scarce attention to the problem, except for some declaration of principles, and did not place sufficient pressure on the central government to pass stringent antismog laws.

CHAPTER 6

TRANSPORTATION AND THE ENVIRONMENT IN ITALY

1944-2006

During the second half of the twentieth century the development of trans-portation in Italy was characterized by three factors: the achievement of road transportation, the construction of great infrastructure (first and foremost highways), and the rapid growth in volume of both passenger and freight transportation.

THE GROWTH OF TRANSPORTATION

By the end of the Second World War the Italian road network presented complex structural problems, essentially war damages which had made half of the state road system, one third of provincial roads, and one sixth of local roads unusable. The road network still traced the pattern of the ancient Roman and medieval roads, and over two-thirds consisted of provincial and local narrow and unpaved carriageways, crossing all towns and villages en route. This lack of infrastructure prompted the Bonomi government to reorganize the transportation sector, which became one of the first to be administratively rationalized with the extinction of the Ministry of Communications and the creation in December 1944 of the Ministry of Transportation.

POLICIES

Thus a two-headed system was born which had in the Ministry of Public Works its management center for the coordination of road transport and in the Ministry of Transport the institutional hub of the rail and air sectors. This dual system meant that the different networks developed independently almost as if they didn't have to interact in the same territory, and also because of the prevalence of the conviction that infrastructure did not significantly influence the economic development, transportation policy developed without serious and effective planning. During the first parliament (1948–1953), the seventh De Gasperi government introduced a bill in 1951 (signed by Salvatore Aldisio, the minister of Public Works) granting to the National Road Board (ANAS) special funds, 40 billion liras, for the improvement of the road system in southern Italy. The Aldisio bill was the first turning point in favor of road infrastructure development, a choice also influenced by the model of economic expansion which was developing in the early 1950s and regarded the automobile industry as a strategic sector to be supported and encouraged.

Within this scenario the highway policy began to emerge, endorsed by two crucial organizations: on June 16, 1952, the Federazione Italiana della Strada (Italian Federation of the Road Transportation), which declared its objective to fund road infrastructure; and the Gruppo Parlamentare Amici dell'Automobile (Parliamentary Group of Car Friends), chaired by the Christian Democrat Salvatore Foderaro. The Amici dell'Automobile was organized into nine working commissions, staffed by 19 senators and 120 members of the Chamber of Deputies (Christian Democrats, liberals, Democratic Socialists, monarchists, neofascists, and some members of the mixed group, but not communists or socialists).

The first highway plan was presented by Giuseppe Romita, the minister of Public Works and one of the exceptional representatives of the Amici dell'Automobile, on November 9, 1954 (Law No. 463, May 23, 1955). The plan provided 120 billion lira, 20 billion of which was designated to the construction and maintenance of ordinary roads in the south, and 100 billion to the creation of the following highways: Milan–Naples, Serravalle Scrivia–Milan, Savona–Ceva, Turin–Ivrea, Brescia–Padua, and the expansion of the Padua–Mestre, Naples–Pompei, and Naples–Bari highways. The Romita plan was practically unopposed. The parties of the left, and the communists and socialists in particular, did not oppose it for fear of damaging the interests of the unemployed. Between 1956 and 1958 center governments more precisely specified their highway policy. In November 1956 Law No. 1328 was passed, which included in the program the Bologna–Rimini highway and the doubling of the Via Aurelia between Rome and Livorno and the Florence–Tyrrhenian Sea and the Naples–Reggio Calabria highways. When the second parliament came to an end on June 30, 1958, the attitude in favor of road transportation was definitely in place. In the course of the five-year term, governments had financed the building of 10,000 km of roads, started the improvement of the national road network, and approved a highway plan which had no equal in Europe.

This attitude continued three years later with the highway plan proposed by Prime Minister Amintore Fanfani on January 12, 1961. The main provisions of the measure (Law No. 729/1961) concerned the construction of some highways, the consent of ANAS to take out loans for 180 billion liras for the construction of the Salerno–Reggio Calabria highway, and the allocation of 408 billion liras (a figure which committed the budget of ANAS until the year 1999–2000) for cofinancing the highways previously approved. In essence the Fanfani plan brought to fruition an infrastructure policy centered on a highway system that was unprecedented in Europe, which resulted in the waning of interest in public transportation. Railways entered a crisis phase caused by aging lines and the difficulty of competing with road transportation, which was favored by a widespread proliferation of buses and because of the flexibility cars granted. The tramways, both urban and suburban, were gradually replaced by buses, and air travel began to play a growing, albeit still marginal, role. Transportation policy, embedded within a two-headed development system (on the one hand roads and cars and on the other railways), ended up promoting micro-sectorial interventions instead of broader integration and duplicated the noninterference and noncommunicability among different sectors of the organization identified as the main problem of the Italian transportation system.

The result was a system characterized by a lack of coordination between the different forms of transportation and by an institutional weakness of public policies, due to the pressures exerted by industrial oligopolies: the car manufacturers, led by Fiat, and the electro-mechanical industries, almost entirely state-owned within the Institute for Industrial Reconstruction group. In the transportation sector, the networks among industry, politics, and public administration tended to merge so as to structure stable and lasting policies which extended to the construction industry (fragmented into many small units) and the petrochemical industry (gathered around the National Hydrocarbon Corporation [ENI]). This closed and protected market, practically unchanging until the early 1990s, generated a mono-modal system characterized by a predominance of road transportation.

In the early twenty-first century the main problem of the transportation system remained the absence of a comprehensive policy that was capable of treating its major chronic woes: the lack of coordination between the different modes of transportation and the existing inequalities among regions. Inter-

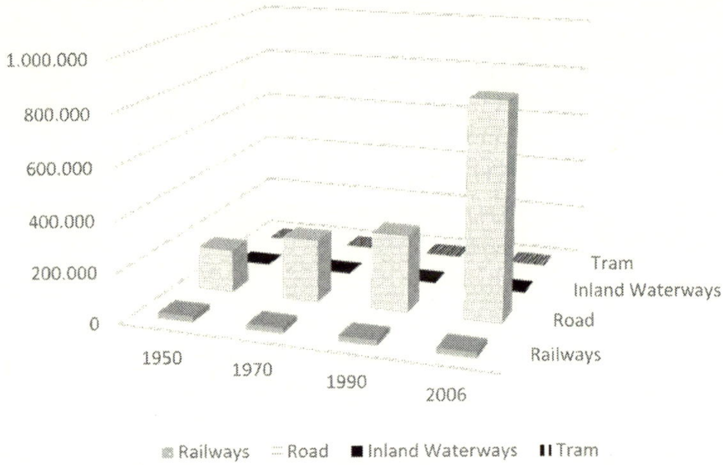

Graph 6.1. The length of infrastructure (km) in Italy, 1950–2006. *Source*: Ministero dei Trasporti, *I trasporti in Italia* (Rome, 1995); ISTAT, *Statistiche ambientali 2009* (Rome, 2009).

modality development was hampered by at least four factors: the predominance of road over rail, the prevalence of small companies over large ones in road transportation, the absence of policies to discourage long-distance trucking, and the underestimation of diseconomies produced by road transportation (e.g., pollution and the overuse of nonrenewable resources).

The persistence of such problems highlights the difficulties encountered by policy makers in putting into practice the directives approved by the European Union (until 1992, the European Economic Community). Despite the many actions favoring rail, water, and sea transportation (which are less detrimental to the environment and less energy-consumptive) and reducing the competitive inequality among the different means of transportation by placing their externalities (including environmental ones) on those who utilize the systems and on communities, the directives were unable to affect this mono-modal nature of Italian transportation.[1]

THE DATA

The mono-modal nature of the Italian system of transportation also appears evident from analysis of the data:[2] between 1950 and 2006, the road network grew by 395.12%, whereas that of the railways and tramways decreased, respectively, by 6.32% and 72.93% (Graphs 6.1 and 6.2);[3] in 2006, roads, which in 1970 moved 44.29% of goods, attracted 64.30% of the traffic (Graph 6.3).

In fairness, Italy is not the only country to have a transportation system decidedly in favor of roads; over the years mono-modality—presented in the

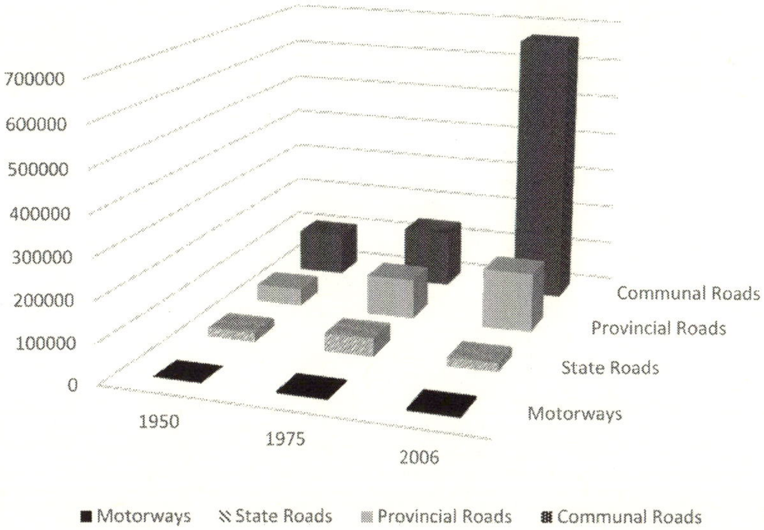

Graph 6.2. The length of Italy's road network (km), 1950–2006. *Source*: ISTAT, *Annuario statistico italiano* (stated years); ISTAT, *Statistiche ambientali 2009* (Rome, 2009).

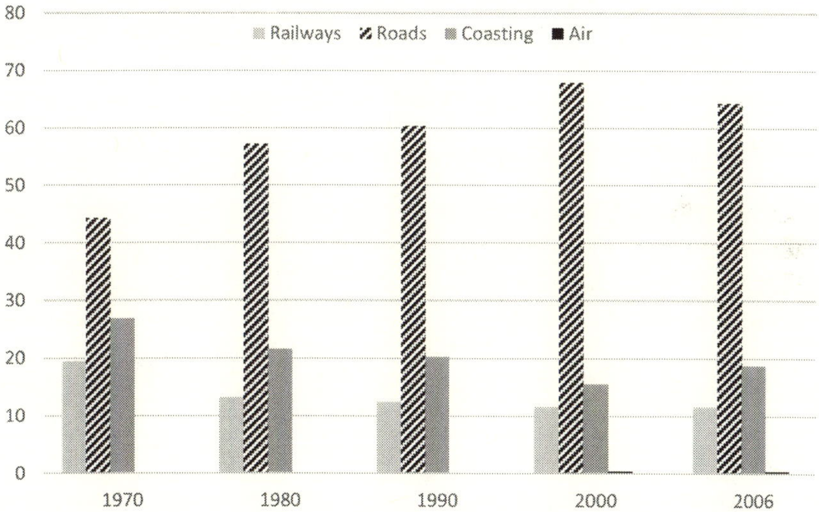

Graph 6.3. Italian freight transportation by volume (percentage of tkm)for medium to long distances, 1970–2006. *Source*: Ministero dei Trasporti, *I trasporti in Italia* (Rome, 1995); ISTAT, *Annuario statistico italiano* (stated years).

historiography as a negative peculiarity of the Italian system—has become a problem of European magnitude. Between 1970 and 2001 road transportation peremptorily established itself in twelve of the fifteen EU member states.

Table 6.1. The modal split of passenger and freight transportation (percentage of pkm and tkm) for EU-15, 1970–2001.

	Passenger transport					Freight transport			
	Car	Bus	Tram + Metro	Rail	Air	Road	Rail	Inland waterways	Oil pipelines
1970	73.8	12.7	1.6	10.4	1.6	52.1	30.2	10.9	6.8
1980	76.1	11.8	1.2	8.4	2.5	60.0	24.1	8.8	7.1
1990	79.0	9.3	1.0	6.7	4.0	69.3	18.1	7.6	5.0
1995	79.5	8.7	0.9	6.2	4.6	73.3	14.1	7.3	5.3
2001	78.2	8.6	1.0	6.4	5.9	75.5	13.1	6.8	4.7

Source: European Commission, Directorate General for Energy and Transport, *Energy and Transport in Figures* (Bruxelles, 2003).

Table 6.2. Freight transportation (percentage of tkm) for EU-15, 2001.

	Road	Rail	Inland waterways	Oil pipelines
Austria	50.4	30.8	4.5	14.3
Belgium	71.0	12.6	13.5	2.8
Denmark	69.5	14.7	12.8	3.1
Finland	72.6	25.9	1.4	-
France	77.8	14.3	1.9	6.0
Germany	69.5	14.7	12.8	3.1
Greece	98.0	2.0	-	-
Ireland	93.3	6.7	-	-
Italy	87.9	8.1	0.1	3.9
Luxembourg	71.2	17.1	11.7	-
Netherlands	46.6	4.0	43.4	6.0
Portugal	87.2	12.8	-	-
United Kingdom	83.4	10.3	0.1	6.1
Spain	87.6	7.5	-	4.8
Sweden	60.5	39.5	-	-
EU 15	75.5	13.1	6.8	4.7

Source: European Commission, Directorate General for Energy and Transport, *Energy and Transport in Figures* (Bruxelles, 2003).

In 1970, 52.1% of all goods traveled by road, 30.2% by rail, 10.9% on inland waterways, and 6.8% in pipelines. In 2001 the extraordinary growth of road transportation became apparent: it consisted of 75.5% of all traffic, compared to 13% for railways, 6.8% for inland waterways, and 4.7% for oil pipelines (Table 6.1).

In 2001, only in Sweden, Austria, and Finland did the railways continue to play a primary role (Table 6.2). The volume of traffic handled by railways

Table 6.3. Passenger transportation (percentage of pkm) for EU-15, 2001.

	Car	Bus	Rail	Tram + Metro	Motorbike
Austria	72.9	13.9	8.6	2.9	1.7
Belgium	82.5	9.5	6.1	0.7	1.1
Denmark	79.3	12.2	7.5	-	1.0
Finland	82.2	11.1	4.7	0.7	1.3
France	84.1	5.1	8.3	1.2	1.4
Germany	80.8	7.9	8.5	1.0	1.9
Greece	64.9	17.5	1.3	1.1	15.3
Ireland	81.0	14.5	3.5	-	0.9
Italy	75.5	10.9	5.3	0.6	7.7
Luxembourg	79.9	13.9	5.3	-	0.9
Netherlands	82.9	6.9	7.9	0.8	1.5
Portugal	79.3	10.7	3.3	0.5	6.2
United Kingdom	86.4	6.4	5.4	1.1	0.7
Spain	76.9	13.0	5.2	1.3	3.6
Sweden	80.3	9.7	7.4	1.9	0.7
EU 15	80.4	8.8	6.5	1.0	3.2

Source: European Commission, Directorate General for Energy and Transport, *Energy and Transport in Figures* (Bruxelles, 2003).

ranked above 10% in only six other countries: Luxembourg, Germany, France, Portugal, Belgium, and Great Britain. In the remaining six it was marginal. Waterways played a significant role only in the Netherlands, Belgium, Germany, and Luxembourg.

The road mono-modality is further highlighted by passenger transportation (Table 6.3). In 2001 private transportation, on four or two wheels, was dominant in all European countries (83.6% of the total, compared to 16.3% for public transportation). The close relationship between the means of transportation and income level appeared evident: in spite of the existence of efficient and widespread public transportation networks, car use reached its maximum in countries with a high per capita GDP. The use of public transportation was lower than the European average in countries with a high level of infrastructure, such as in France, the Netherlands, and the United Kingdom (Finland and Germany ranked a little above the European average). The use of two-wheeled vehicles was strictly connected with the average per capita GDP and the scarcity of public transportation in metropolitan areas.[4]

The situation doesn't change when we consider the twenty-seven EU member states: in 2006 passenger and freight transportation on roads remained dominant (Graph 6.4).

THE ENVIRONMENTAL IMPACT OF TRANSPORTATION: PRODUCTION AND INFRASTRUCTURE

Throughout the twentieth century transportation has been one of the main agents of environmental change. The interactions between transportation and the environment are much more complex than are acknowledged today in the debate between policy makers and social actors, which is chiefly centered on the effects produced by its daily use.[5]

The impact of transportation on the environment manifested itself following the construction of industrial plants. Beginning in the early years of the twentieth century, factories have profoundly altered the environment of the places where they have been established: land consumption (built in the 1930s, Fiat Mirafiori alone occupied 2 million m^2); repercussions on the physiognomy of urban development and architecture (built in what at the time seemed peripheral areas, far from residential quarters, industrial plants have been swiftly enveloped by the urban network, following intense development that was fueled by immigration);[6] social implications (tensions generated by migratory flows); problems raised by the restoration of industrial areas as the dynamics of the economy required the restructuring of production cycles (consider the case of the industrial desert of Lambrate where the production area of Innocenti was located, for years the center of controversy fueled by assumptions about its future use).

The impact on the land of the production of transportation means goes well beyond the physical places where the factories are. The construction of the infrastructure needed to make transportation functional has resulted in a huge consumption of space:[7] at the beginning of the twenty-first century roads occupied nearly 5,914,804 m^2, railways 126,630 m^2, airports 145,940,000 m^2, ports (yards for the handling of goods only) 11,034,012 m^2.

Until recent years infrastructure was designed and built without taking into consideration the effects on the landscape and natural resources.[8] As Massimo Moraglio has written, "[Everything] was aimed at emphasizing the effectiveness . . . of human activity, its power in modifying the land, in crossing rivers, in tunneling through mountains and in crossing valleys to reduce distances and facilitate communication."[9] Thus the values of landscape and natural resources became obscured by the "infrastructural epic" which enjoyed a broad approval (both from the ruling class and among citizens) because it was a tangible demonstration of modern Italy, of a country which had left behind a past of poverty and misery in order to enter the club of the most

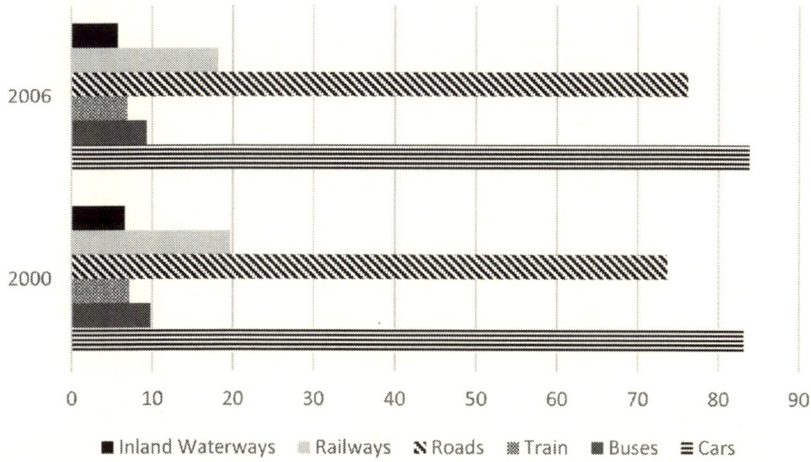

Graph 6.4. The modal split of passenger and freight transportation (percentage of pkm and tkm) for EU-27, 2000 and 2006. *Source*: Eurostat, tables interface online (Nov. 17, 2011), https://ec.europa.eu/eurostat/data/database (no longer available).

industrialized countries. The tracks of high-speed rail lines, airports, and highways not only served as means of transportation, but also demonstrated to the countries of the First World that Italy was about to join them, entering through the main door (of accelerated modernization resting on the pillars of industrialization and mass consumerism).

The long wave of environmental impact caused by the transportation sector on the land does not stop here: transportation has represented the prime flywheel for the development of two highly polluting industries: oil and rubber (Table 6.4). In 1993, of thirteen areas declared to be under "high risk of environmental crisis," five hosted petrochemical industries and refineries (Naples, Brindisi, Taranto, Priolo-Agusta, and Gela). In 2006, 52.9% of those industrial plants at risk of significant accidents belonged to the chemical, petrochemical, and gasoline refining sectors.

There is one highly important issue regarding the flow of materials. A country's economy may be regarded as a kind of large metabolic organism regulated by the flow of materials to and from it: among the latter are the emissions and refuse generated by the production and consumption of goods.[10] Industrial activities (like all other human activities) end up creating a sort of degenerative metabolism by which industries use natural resources (e.g., water, energy sources, soil) and return them in the form of degraded metabolites (e.g., emissions into the atmosphere and water, waste deposited at dumps). Part of the exit flows consists of the product of *external diseconomies*

Table 6.4. Industries in Italy that produce the byproducts of oil and rubber (1,000 tons).

	1980	1985	1990	1995	2000	2004
Car petrol	15,218	14,665	18,011	18,723	20,875	20,759
Aviation fuel	193	84	48	26	9	2
Jetfuel	1,840	1,364	2,172	2,164	2,753	2,551
Petrol	2,419	2,190	--	1,727	1,607	1,408
Diesel oil	25,215	22,712	28,640	31,295	33,891	38,025
Fuel oil	35,996	20,066	21,952	17,281	14,349	13,278
Lubricating oils	1,028	974	1,172	1,292	1,167	1,239
Oil bitumen	2,065	2,064	2,356	2,330	2,686	3,496
Liquified gas	2,027	1,742	2,154	2,235	2,308	2,398
Virgin nafta	2,637	2,596	1,806	3,466	3,448	3,939
Light distilled fuels	173	152	10	6	--	--
Pneumatic tyres	355	306	324	348	396	279

Source: ISTAT, *Statistiche ambientali* (Rome, 1991, 1993, 2002, 2007).

(externalities) generated by industrial activity.[11] Transportation is the cause of *unidirectional* externalities, for example, the exploitation of water resources and the emission of pollutants whose costs are imposed on the community in the form of damages to health and cleanup activities that do not enter into the economic calculation of the producing firms; *reciprocal* externalities (such as emissions from vehicle exhaust), perceived as a social damage produced by a collective behavior whose magnitude is regarded as very modest per individual since it derives from the use of the environment by numerous members of the community; and *intertemporal* externalites, such as the emissions of gas impacting future generations.

The limited and fragmentary data available do not provide a comprehensive picture of the environmental impacts caused by the metabolism (flows of matter and externalities) of the production of transportation.[12] However, it is possible to construct a partial synthesis using a few indicators: water unit per capita loading (equivalent population),[13] the concept of *virtual water*, and the average quantity of waste produced by each worker.

By analyzing the indicator of equivalent population, it is clear that, with respect to water resources, the transportation industry does not have a high pollution load: in 1981 it amounted to 702,649 units (0.53% of the total equivalent population), and in 2000 it had decreased to 459,775.[14] Interestingly, in order to assess the externalities of the production of transportation, one must consider the sources of water extraction and the discharge bodies used. In 1990 the sources for extraction were: 64.3% wells, 18.5% waterworks, 17% other sources, 0.1% sea, and 0.1% lakes.[15] This highlights two problems that gripped Italian industry, and which evidently did not escape even the transportation

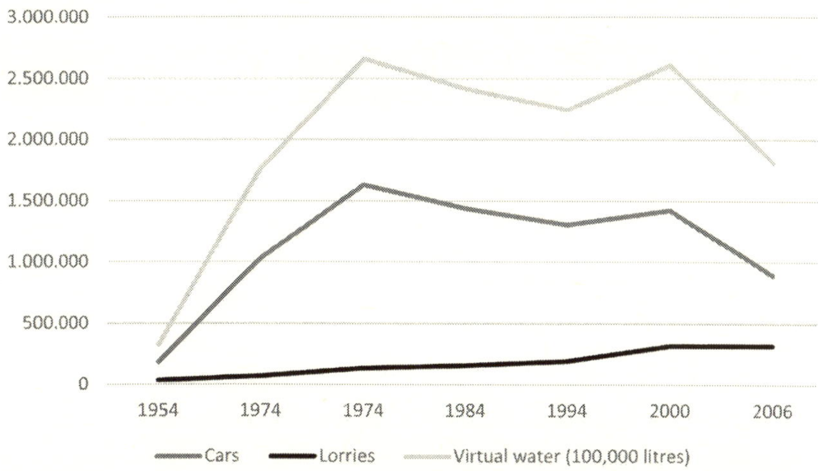

Graph 6.5. The estimate of virtual water in the production of Italian cars and trucks, 1954–2006. *Source*: Author's estimates according to the method indicated in chapter 6, note 18.

manufacturers: the dependence on groundwater, which was causing severe piezometric depressions in industrial areas; and the increasing reliance on surface water—prompted by the subsidence of the water table which prevented an increased use of wells—by municipalities for their waterworks. Industrial use of water provided by municipal waterworks caused severe conflicts over this resource: the scarcity of water in summer months generated a true competition for water resources among the municipalities, agricultural consortia, and industries, all equally convinced that they possessed a kind of *priority right* to that water.

Discharge water flowing into sewers accounted for 58.7% of the drainage (33.3% private and 25.4% public), while 25.2% flowed into streams, 14.8% into other places, 1.2% into the sea, and 0.1% into lakes.[16] These figures also highlight another problem: the lack of adequate sewage treatment plants. In the early 1990s, compared to an industrial population equivalent need for more than 92 million units, sewage treatment plants had a capacity of just over 15 million.

A further indicator for the use of water resources—and very effective for illustrating the pressure imposed by industrial activities on natural resources—is represented by the concept of *virtual water*, or water incorporated within the production and trading of foods and consumer goods (this water is hidden within the goods and employed for their production and trading).[17] Graph 6.5 shows the close relationship between industrial production and the total volume of water utilized in the production of motor vehicles.[18] Still

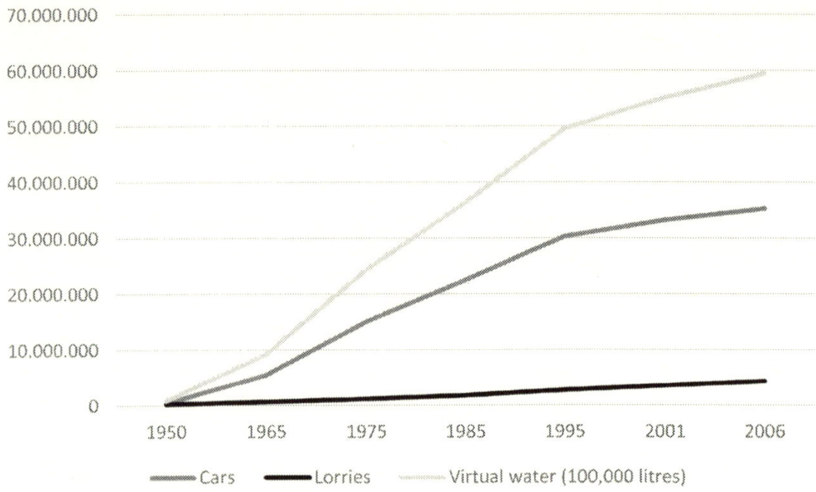

Graph 6.6. Vehicles in use in Italy and estimates of their virtual water, 1950–2006. *Source*: Author's estimates according to the method indicated in chapter 6, note 18.

more effective for understanding the relationship between vehicles and use of natural resources is the volume of virtual water of vehicles traveling on the roads (Graph 6.6). Graph 6.5 enables us to introduce another concept: the *water footprint*. Each country possesses an *internal* water footprint (the amount of water necessary for the production of goods and services produced and consumed within the country) and an *external* one (the water used in the production of goods for export). The water footprint of production is made up of *green water* (the volume of rainwater stored in the soil and subject to evaporation), *blue water* (freshwater pumped from water bodies which is used and not returned), and *gray water* (the volume of polluted water issuing from the production processes). Considering that in 2001, 60% of cars in circulation were manufactured abroad (historically 1.48% in 1958 and 16.58% in 1970), this means that Italy also imports a portion of the exporting countries' water footprint. In 2001, therefore, 60% of the virtual water used in cars currently in circulation came from the water reserves of other countries. This provides a measure of the progressive globalization of environmental problems: the purchase and use of a car no longer affects only the natural resources of Italy, but also those of exporting countries. Seeing this in a global perspective, the improvement shown by indicators in Italy (due to the decrease in production volumes) has a positive effect on the national environment (less consumption of green and blue waters and diminished discharges of gray waters), but at the cost of imposing a part of Italy's domestic environmental problems on producer countries. So the problem of exploitation of natural

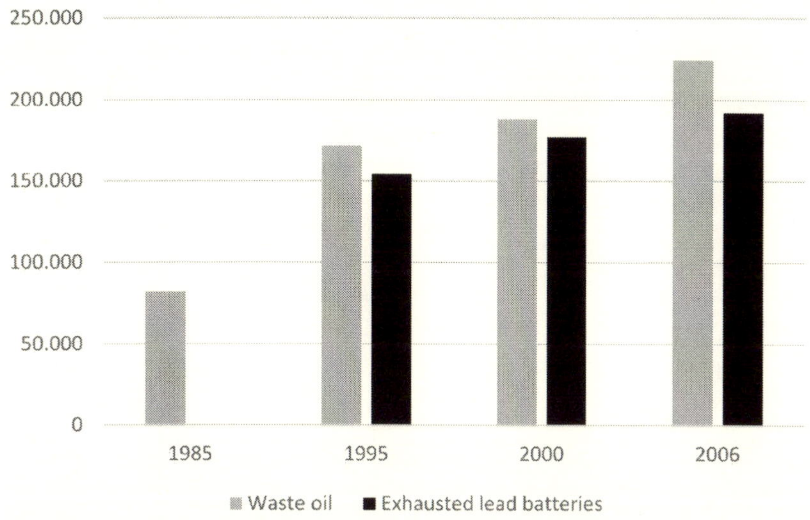

Graph 6.7. Collection of waste oil and exhausted lead batteries (tons), 1985–2006. *Source*: Ministero delle Infrastrutture e dei Trasporti, *Conto nazionale delle infrastrutture e dei trasporti: Anno 2008* (Rome, 2008).

resources does not diminish, but shifts from consumer countries to producer ones.

Let us finally reflect on the third indicator: the average amount of waste per worker. In 1986 a worker in the construction and assembly of motor vehicles produced 448 kg of waste that is similar to urban waste, 829 kg of special waste, and 39 kg of toxic and hazardous waste. Tabulating the total number of workers, the industry as a whole generated 122,730,944 kg of waste similar to urban waste, 227,107,037 kg of special waste, and 10,684,167 kg of toxic and hazardous waste. That same year a single worker in the construction of transportation means, aside from motor vehicles, produced 159 kg of waste similar to urban waste (22,159,830 kg in total), 2,123 kg of special waste (295,882,510 kg), and 13 kg of toxic and hazardous waste (1,811,810 kg).

In 1999 the production of transportation means generated 1,204,515 tons of non-dangerous special waste and 141,224 of hazardous waste. The magnitude of the waste problem is also extrapolated from data on the recovery of some materials (Graph 6.7). Until the mid-1970s a significant part of industrial waste was directly dumped into water, giving rise to serious pollution incidents. The improper disposal of waste products was a major cause of soil contamination as well. This was spread in small amounts everywhere, with hundreds of sites responsible for the storage of sewage and industrial slag. Since the mid-1980s the situation has gradually improved thanks to a legislative intervention for the classification of waste disposal on July 27, 1984, which, among other provisions, favored the recovery of certain materials. But still today waste deriving from production processes that dissipate large amounts of materials and energy represents a significant environmental externality.

THE ENVIRONMENTAL IMPACT OF TRANSPORTATION: CONSUMPTION AND UTILIZATION

In the course of the second part of the twentieth century two trends have become clear: the rapid growth in car use and burgeoning air transportation (Tables 6.5 and 6.6). In 1990 Italy had become the leading European country in car density: there were 483 cars per one thousand residents, compared to 480 in Luxembourg, 447 in Germany, 421 in Sweden, 415 in France, 389 in Finland, 388 in Belgium, 387 in Austria, 368 in the Netherlands, 360 in Great Britain, 309 in Denmark and Spain, 227 in Ireland, 187 in Portugal, and 171 in Greece.

The high level of Italian car ownership was characterized by uninterrupted growth: cars in circulation grew from 15,059,689 in 1975 to 27,415,828 in 1990. Sixteen years later 46,329,144 vehicles traveled on Italian roads, among which were 4,331,747 trucks and 5,288,818 motorcycles. The latter returned to being the principal alternative to the car for traveling on the city roads, which had grown more and more congested. With the exception of Luxembourg (280,709 cars for 444,100 residents), Italy established itself as the leading European country in car density, with one car for every 1.72 residents, compared to 1.96 in Germany, 2.03 in Austria, 2.04 in France, 2.15 in Belgium, 2.21 in Sweden, 2.30 in Spain, 2.34 in Portugal, 2.39 in Finland, 2.60 in the Netherlands, 2.61 in Great Britain, 2.78 in Denmark, 2.86 in Ireland, and 3.32 in Greece. In 2006 Italy, with 2,323,440 new registrations, was the third-largest car market in Europe, after Germany (3,467,961) and Great Britain (2,344,864).

Air transportation also grew: in 2001 there were 1,238,239 airplanes (+81% compared to 1992) which transported 90,210,038 passengers (+76%). In 2006 total internal passenger traffic (13,930 mio pkm) had more than doubled compared to 1998 (6,416 mio pkm).

The rapid growth in private car ownership has therefore represented the principal environmental impact factor: the car turned out to be a voracious consumer of space and energy and has produced significant effects on land use (and urban form), on energy consumption, and on the volume of polluting emissions. Affecting land use and the consequent transformation of the urban form, local administrations between the 1950s and the 1980s did not make choices capable of adapting urban planning to the increased demand for urban mobility.

Table 6.5. The prevalence of passenger transportation by type (1,000 million pkm) for EU-15, 1970–1998.

	Powered 2 wheelers	Car	Bus and Coach	Tram Metro	Rail	Sea	Air
1970	113	1,586	280	39	217	15	33
1980	127	2,307	364	41	253	22	74
1990	126	3,253	385	48	274	28	157
1998	134	3,814	416	50	290	32	241
1970-1998	+18.6%	+140.5	+48.6%	+28.2%	+33.6%	+113.3	+630.3%

Source: European Union, *Transport and Environment: Data 1980–1999* (Luxembourg, 2001).

Table 6.6. The prevalence of passenger transportation by type (1000 mio pkm) for EU-27, 1999–2006.

	Powered 2 wheelers	Car	Bus and Coach	Tram Metro	Rail	Sea	Air
1999	133	4,240	515	75	359	43	425
2002	138	4,480	519	78	366	42	445
2006	152	4,680	528	84	389	40	549
1999-2006	+14.3%	+10.4	+2.5	+12.0	+8.4	-4.6	+29.2%

Source: European Union, *Energy and Transport in Figures* (Luxembourg, 2010).

There are two reasons explaining such a decision-making deficit. The first concerned the absence of suitable legislation, which had remained unchanged since 1942 because Parliament had been unable to pass new general urban planning laws. The second concerned the approval of a law (No. 291/1971) which established anti–economic crisis measures in favor of the building trades. The law helped to deregulate further town planning, aggravating the instability of urban areas, and exempted municipalities from adhering to minimal planning standards. This had the effect of invigorating building speculation, which in the absence of an effective legislation detracted from planning ever-increasing portions of public space and prevented their future use for infrastructural adaptation and the expansion of collective spaces (e.g., public greenspaces and infrastructure).

The resulting expansion of Italian cities has caused the negative traits typical of *urban sprawl*, a horizontal growth pattern which occupies considerable land, making cars essential. Urban sprawl is most prominent in countries and regions with a high population density and intensive economic activity (in Italy the north and the industrial areas of the central regions). The planning policies at both the local and regional levels—fueled by EU cohesion and structural funds to support infrastructure developments—have ended up accelerating urban sprawl in response to improved transportation links and enhanced personal mobility, and has caused the emergence of new develop-

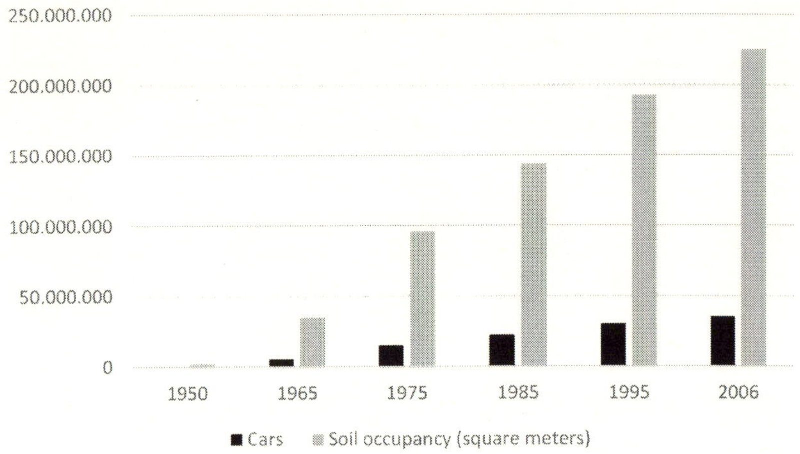

Graph 6.8. Cars in use and estimates of land (soil) occupancy, 1950–2006. *Source*: Author's estimates according to the method indicated in chapter 6, note 19.

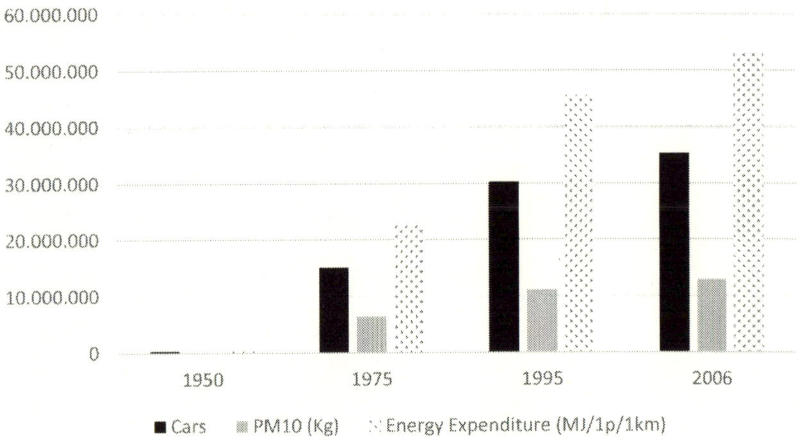

Graph 6.9. Cars in use and estimates of PM_{10} emissions and energy expenditure, 1950–2006. *Source*: Author's estimates according to the method indicated in chapter 6, note 22.

ment patterns around the smaller cities situated along transportation corridors. This tendency is causing an increase in land consumption: the amount of space consumed per person in the cities of Europe has more than doubled over the past fifty years; over the past twenty the extent of built-up areas in many Western and Eastern European countries has increased by 20% while the population has increased by only 6%.

A close connection exists between urban expansion and environmental effects: the urban sprawl causes an increase in the demand for private mobility,

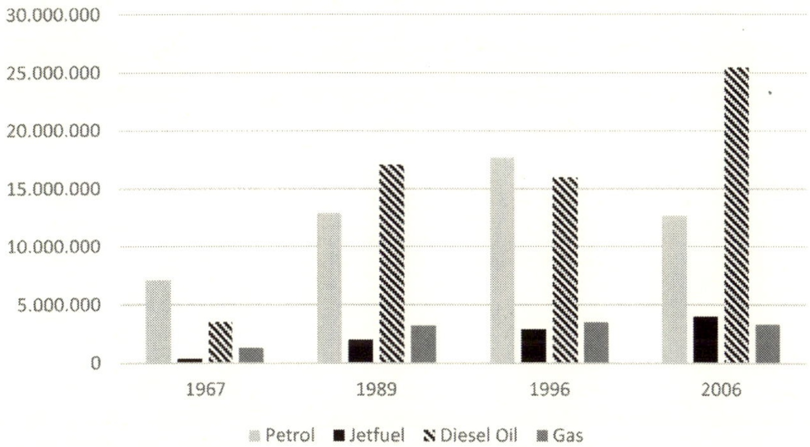

Graph 6.10. Consumption of oil products (tons), 1967–2006. *Source*: ANFIA, *Automobile in cifre* (Turin, 1971); ISTAT, *Statistiche ambientali* (Rome, 1998, 2008).

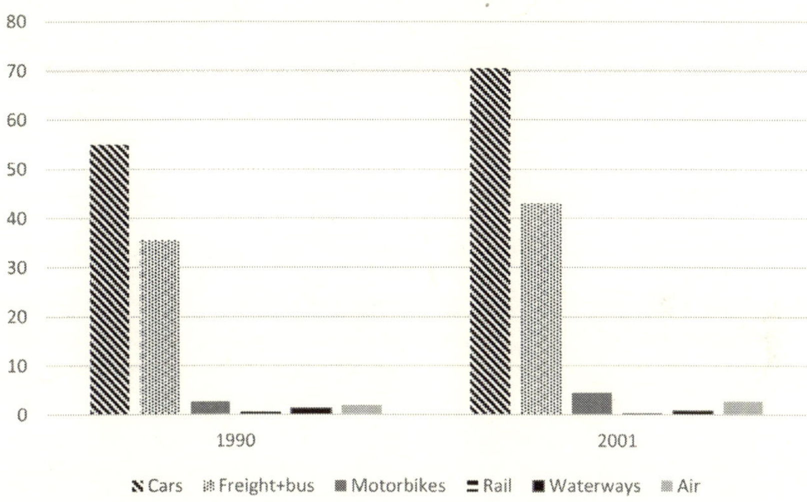

Graph 6.11. Carbon dioxide emissions (1,000 tons) from transportation in Italy, 1990 and 2001. *Source*: Author's estimates on the basis of ISTAT and ENEA data.

which in its turn generates a further consumption of space (such as the space occupied by cars; see Graph 6.8)[19] and increases pollution-related health problems.[20] According to a report on air pollution in eight major Italian cities compiled by the WHO, PM_{10} (fine particulate matter) alone is responsible for 3,500 deaths, 1,900 hospitalizations for respiratory problems, 2,700 hospitalizations for cardiovascular diseases, and 31,500 acute bronchitis.[21] The growth of private car ownership also had impacts on energy consumption or on the volume

Table 6.7. Emissions (tons) from transportation in Italy, 2005.

	SOx	NOx	NMVOC	CH_4	CO	CO_2	N_2O	NH_3
Cars	1.306	226,766	122.202	16.443	1,430.309	70,218,400	10,335	14,980
Light lorries	341	71,203	9,503	544	78,320	14,597,013	972	278
Heavy lorries/ buses	673	195,303	32.439	1.860	62.606	28.109,352	1,174	117
Powered 2 wheelers	65	5.689	127,037	8,320	542,516	4.084,090	71	71
Evaporation	--	--	75,647	--	--	--	--	--
Road Transport	2.385	498.961	366,828	27,167	2,113,751	117,008,855	12,552	15,446
% Cars on road total	**54.76%**	**44.45%**	**33.31%**	**60.53%**	**67.67%**	**60.01%**	**82.34%**	**96.98%**
Railways	7	3,841	451	17	1.038	302,826	120	1
Inland waterways	2	1,052	117	4	270	77,252	2	--
Coasting	57,831	88,653	112,513	1,534	259,991	6,799,664	158	14
Air	847	11,926	781	83	3,273	2,651,671	62	--
% Road on total	**3.91%**	**82.55%**	**76.31%**	**94.31%**	**88.87%**	**92.25%**	**97.34%**	**99.90%**

Source: ISTAT, *Statistiche ambientali 2008* (Rome, 2008).

of polluting emissions (Graphs 6.9 and 6.10).[22] Cars have played a dominant role in the production of carbon dioxide (Graph 6.11), and road transportation alone is the main source of sulfur oxides, nitrogen oxides, methane, carbon monoxide, carbon dioxide, nitrogen protoxide, and ammonia (Table 6.7).

Analyzing the relationship between the consumption of transportation and the onset of critical environmental issues from a European perspective, we can underline the energy-intensive characteristics of transportation and the close relationship existing between traffic volumes and emissions. The problem appears related to the consolidation of a very specific social model of mobility characterized by high demands for an ever-faster and widespread mobility in travel by private vehicles and public transportation that use fossil fuels. In this context, the remarkable voracity for energy and the high volume of emissions are common traits in all countries of the European Union (Tables 6.8 and 6.9). True, some countries are more virtuous than others, but the trend is very similar in all countries of the European Union (particularly those belonging to the EU-15).

* * *

In the second half of the twentieth century, the growth of transportation has been one of the hallmarks of industrialized societies (by the end of the century this included developing countries). This development has been characterized by the consolidation of road transportation, by the construction of major infrastructure (primarily highways), by rapid growth in the volume of passenger and freight transportation, by the spread of personal mobility for domestic routes, on two or four wheels, and air transportation for international and transoceanic routes.

At the same time transportation has been one of the principal causes of

Table 6.8. Final energy consumption of transportation in EU-15 (1985, 1998) and EU-27 (2006).

	1985		1998		2006	
	Mtoe	%	Mtoe	%	Mtoe	%
Austria	4.5	23.5	6.3	27.5	7.7	28.7
Belgium	8.4	26.8	15.0	35.0	9.6	25.1
Denmark	4.0	27.5	6.1	37.2	5.3	34.0
Finland	3.8	20.0	4.8	20.0	5.0	18.7
France	35.9	26.9	51.5	33.6	50.9	32.3
Germany	51.6	21.7	66.7	29.6	63.3	28.4
Greece	5.8	42.4	10.8	49.7	8.5	39.5
Ireland	1.7	27.5	3.5	36.8	5.4	41.5
Italy	31.2	31.2	43.6	34.5	44.2	33.8
Luxembourg	0.6	20.2	1.6	48.9	2.6	59.1
Netherlands	17.5	34.1	25.8	42.0	15.6	30.7
Portugal	3.1	31.3	6.1	38.0	7.1	38.4
United Kingdom	38.0	29.3	53.0	35.0	55.9	37.2
Spain	17.7	35.3	36.4	47.2	40.7	41.8
Sweden	7.0	22.0	9.3	26.5	8.6	25.9
EU 15	**230.9**	**27.1**	**340.4**	**34.5**	**330.4**	**34.3**
Bulgaria	--	--	--	--	2.8	28.0
Cyprus	--	--	--	--	9.9	50.0
Czech Republic	--	--	--	--	6.3	24.1
Estonia	--	--	--	--	0.8	28.6
Hungary	--	--	--	--	4.7	26.3
Lithuania	--	--	--	--	1.5	31.9
Latvia	--	--	--	--	1.2	28.6
Malta	--	--	--	--	0.3	60.0
Poland	--	--	--	--	13.4	22.1
Romania	--	--	--	--	4.4	17.8
Slovenia	--	--	--	--	1.6	32.6
Slovak Republic	--	--	--	--	1.8	16.8
New EU Members	--	--	--	--	**48.7**	**30.6**

Source: European Commission, Directorate General for Energy and Transport, *Energy and Transport in Figures* (Bruxelles, 2001, 2009).

environmental change. The impact of the production of transportation means has had significant consequences on the processes of urbanization and on the organization of the urban form, and also on the use of resources (water in particular, as well as energy) and the production of waste. Motor vehicles' daily use has had significant effects on air quality (especially in urban areas where polluting emissions are concentrated), on the demand for oil by-products, and on climatic balances. Private mobility on the road in particular has played a fundamental role in urban expansion (made possible, for the greater part, by the flexibility allowed by two and the four wheels), and it has been a promoting factor in the growth of land consumption (in Italy, between 1950 and 2006,

Table 6.9. Carbon dioxide emissions as a share (%) of total national emissions in EU-15, (1985, 1998) and EU 27 (2006).

	1985		1998		2006	
	Total transport	Road transport	Total transport	Road transport	Total transport	Road transport
Austria	25.0	23.1	30.2	26.9	31.1	27.7
Belgium	18.1	15.3	23.8	18.9	37.4	16.3
Denmark	17.6	13.6	23.3	18.2	30.6	19.8
Finland	21.0	18.3	23.1	19.7	23.7	16.7
France	26.9	24.0	37.4	32.2	38.0	30.0
Germany	14.0	12.1	22.6	19.9	20.9	16.4
Greece	24.5	15.9	25.1	17.6	29.4	16.2
Ireland	19.1	16.2	25.5	20.9	33.1	25.9
Italy	23.8	21.7	28.8	26.2	28.6	23.4
Luxembourg	17.5	15.0	58.7	47.8	61.7	52.6
Netherlands	18.1	15.5	23.5	16.6	42.9	14.6
Portugal	31.3	24.5	32.8	28.5	34.3	27.5
United Kingdom	19.1	15.3	26.8	20.7	29.0	20.2
Spain	24.8	19.7	36.5	29.1	35.8	24.0
Sweden	31.4	27.1	42.3	36.5	48.0	30.5
EU 15	**19.7**	**16.8**	**28.0**	**23.4**	**35.0**	**24.1**
Bulgaria	--	--	--	--	16.9	13.6
Cyprus	--	--	--	--	36.8	20.6
Czech Republic	--	--	--	--	14.4	13.3
Estonia	--	--	--	--	19.0	12.6
Hungary	--	--	--	--	21.2	19.8
Lithuania	--	--	--	--	33.2	27.8
Latvia	--	--	--	--	45.9	34.1
Malta	--	--	--	--	60.1	9.4
Poland	--	--	--	--	11.9	10.8
Romania	--	--	--	--	11.5	10.8
Slovenia	--	--	--	--	28.0	27.1
Slovak Republic	--	--	--	--	14.8	14.2
New EU Members	--	--	--	--	**26.1**	**17.8**

Source: European Commission, Directorate General for Energy and Transport, *Energy and Transport in Figures* (Bruxelles, 2001, 2009).

land occupied by cars has increased from 2,180,179 to 224,998,994 m^2 with a percentage growth of 10,220%).

The Italian case does not present special peculiarities when compared to other EU countries. The perception of a greater gravity of the problem (not only of an environmental nature) induced by transportation, which in the last decades has pervaded the sociopolitical debate in Italy, does not have statistical grounding because it is essentially due to the disordered planning that characterizes Italian cities and has repercussions on the organization and efficiency of public transportation.

The chaotic traffic which grips urban centers in Italy; the lack of efficient public transportation (predominantly buses rather than trams and metropol-

Table 6.10. Greenhouse gas emissions (Mio tons CO_2 equivalent) from transportation in EU-27.

	Air	Road	Railways	Navigation	Other	Total transport	Total emissions
1991	81.7	723.5	13.0	129.2	11.1	958.5	5,644.2
1996	106.5	805.2	10.7	136.6	10.2	1,069.1	5,523.7
2001	133.0	877.1	9.1	157.2	9.2	1,185.6	5,368.8
2006	157.1	924.2	8.3	197.4	10.3	1,297.3	5,447.8
1991-2006	+92.2%	+27.7%	-36.2%	+52.8%	-7.2%	+35.4%	-3.5%

Source: European Commission, Directorate General for Energy and Transport, *Energy and Transport in Figures* (Bruxelles, 2009).

itan railways), which is characterized by exasperating slowness; the irrational organization of urban spaces as a result of the unplanned expansion of cities; and the particular preference Italians have for individual mobility[23] (among countries with a medium-high GDP, Italy is the only country with a significant number of motorcycles) have caused a crisis whose evolution seems to be going around in a circle. An increase in the number of private vehicles led to a worsening of urban traffic and increasingly inefficient public transportation, which led to even worse solutions by municipal administrations (e.g., the extension of parking in central districts and populistic provisions such as the suppression of parking fees), which further increased the number of private vehicles.

All of this happened (and is still happening), albeit on a smaller scale, in cities of other EU countries (with the partial exception of Mediterranean states with low incomes: Greece and Portugal), even where public transportation is more efficient, and municipal administrations are less inclined toward populistic attitudes, and where a collective sensibility is less prone to tolerating the breaking of traffic rules (e.g., random parking in illegal spaces and bicycles and motorcycles traveling in the opposite direction of traffic or on pavements). Such problems have repercussions on the daily lives of citizens (and in consequence, upon their perception of the problems), even where they do not impinge on the environment.

It is important, finally, to emphasize the existing relationship between technology and increased consumption: the replacement of leaded gasoline with unleaded, the improvement of diesel fuels, the use of catalytic converters, and the technological upgrades of engines to the allowable maximum emissions set by European regulations, which has caused a decrease in lead, nitrogen oxide, and sulfur dioxide levels. If the results achieved were not those initially hypothesized—chiefly for the emission of greenhouse gases, which increased by 27.7% between 1991 and 2006 (see Table 6.10)—this is because technological progress has been partially neutralized by the constant increase in the use of motor vehicles (in Italy between 1950 and 2006 the number of cars and trucks in circulation had increased by 6,766%) and the overall increase in all kinds of transportation.

The data relative to emissions highlight that regulating measures implemented to date (European directives imposed within the national legislation concerning the quality of fuels, emissions from motor vehicles, and maximum levels of principal pollutants) have scarcely been effective. In particular, the analysis of data highlights an absence of effective policies aimed at containing road transportation and encouraging local public and rail transportation.

At the beginning of the twenty-first century, therefore, Italy appears to be still incapable of adjusting its policies to the challenges demanded by European integration and globalization due to an apparent inability to make its transportation system more efficient by overcoming the two-headed development model (on the one hand the policies favoring road and car transportation and on the other, rail), the dismantlement of monopolies and protected markets, and the promotion of effective actions aimed at tackling the problem of mobility in urban areas.

CHAPTER 7

CARS AND EMISSIONS REGULATION IN ITALY

1950–2008

The earliest studies linking photochemical smog with the residual products of combustion from motor vehicles were published in the United States in the early 1950s. In 1950 the California biochemist Arie Haagen-Smit showed that there was an undeniable relationship between the smog enveloping Los Angeles and car exhaust; he was able to establish that photochemical smog was produced by the reaction of hydrocarbon compounds with nitrogen oxides under the influence of sunlight. The Air Pollution Control District of Los Angeles concluded that road traffic was responsible for 50% of the air pollution in the United States, estimating that a motor vehicle engine produced 3.37 kg of poisonous substances for every 10 liters of gasoline consumed, and that for damages caused by smog, every US citizen should pay an annual tribute of ten dollars.[1]

AN INCONVENIENT DISCOVERY: THE CAR MAY DAMAGE ITALY'S ENVIRONMENT

In the 1950s the car, along with home heating and industrial plants, began to be suspected of being a major cause of air pollution in Italy as well. In 1952 the problem was reported to the Chamber of Deputies by MP Palenzona who presented a parliamentary question to the minister of Industry and Trade and Industry and to the high commissioner for Public Health. The response of Minister Pietro Campilli was very reassuring: after all, Italy had only 781,992 motor vehicles, and the request by MP Palenzona must have appeared as little more than bizarre.[2]

The problem of air pollution caused by motor vehicles at that time seemed an issue of secondary importance for at least three reasons. First, Italy was far from being a motorized country: car density[3] continued to be very low. In 1952 there was one car for every 92.9 residents, compared, for example, to 55.9 in West Germany, 21.8 in France, 19.9 in Great Britain, 7.5 in Australia, and 3.5 in the United States.[4] Second, technological determinism and technology enthusiasm, which pervaded society in the 1950s, induced Italians to think that scientific and technological progress would be able to solve any eventual problems caused by pollution. Third, environmental protection was considered a luxury which could only be afforded by those enjoying a high standard of living, which is why issues relating to environmental policies were considered much less urgent than those designed to ensure better living standards and a minimum of social security to the large sections of the population who still lived in poverty. The question of atmospheric pollution was also ignored by the car manufacturers, who tended to minimize the concerns raised by scientists and to disseminate information contradicting that collected by independent researchers. In the first half of the 1950s, the car trade was proving a very profitable business with huge market penetration, and the industry was not prepared to see their potential earnings lowered by the cost of equipping cars with emission-control devices that, among other things, were not any stimulus to sales. The car manufacturers were also opposed to the dissemination of data on the harmful effects of exhaust fumes for fear that disclosure of such alarming news would inevitably cause a slowdown in car sales at a time when the conditions for an unprecedented expansion of car ownership were at hand.

In November 1957 a conference on air pollution, Congresso per l'Inquinamento Atmosferico, was held in Milan in which fifty delegates from twenty-three European countries, who had responded to calls from the WHO, took part.[5] For the first time in Italy, cars were described as a source of danger and reports showed that in some areas of Milan the level of carbon monoxide in the air was similar to that of Los Angeles, or equivalent to one part per 4,200. Following the conference in Milan, the National Research Council entrusted research on the *malaria urbana* (urban foul air) to a scientific committee whose conclusions were to be used in the preparation of a series of bills.[6]

In the September-October 1964 issue of the *Annals of Public Health*, the Ministry for Health published a study by Professor Cominelli which unequivocally identified road traffic as a major source of pollution.[7] In June 1964, having been urged by the Council of Europe's Consultive Body since 1961, a European conference on air pollution was held, during which it was decided to create a committee of experts which became operational only in June 1966. Its activities included the monitoring and analysis of pollutants, the collection of statistical data, and research aimed at comparing national legislations.

THE FIRST LAW ON AIR POLLUTION AND THE MALARIA URBANA OF CITIES

In this climate, on September 29, 1965, the Senate Health Standing Committee began a hearing on the first bill aimed at controlling air pollution from motor vehicles.[8] The bill consisted of some very general statements: that motor vehicles were not to "produce polluting effluents, however caused" and that diesel vehicles were not to "emit fumes of an opacity above the levels set by the regulation." The only concrete measure provided in the bill stated that the periodic maintenance of these vehicles should include the monitoring of pollutants. The bill stipulated that the Ministry for Health was empowered to make the installation of devices on motor vehicles to reduce the toxicity of exhaust gases mandatory only after asking for a long series of opinions. The law (No. 615, July 13, 1966) was severely criticized by Italian experts, since any reference to planning and technical tools to seriously limit the emission of poisonous substances was absent.

Although Italy was not regarded as being in line with air pollution legislation that existed in other countries with high motor vehicle densities, in reality the rules approved by Law No. 615 were similar to those passed in other European countries. The main difference was that Law No. 615 did not require any real action capable of lowering the levels of air pollution because the Italian Parliament did not approve rules for the effective enforcement of the law.[9]

The most advanced legislation undoubtedly came from California: in 1963 this American state recommended the adoption of exhaust gas recirculation and in 1966 approved the first rules limiting car exhaust: the concentration of carbon monoxide could not exceed 1.5% and that of hydrocarbons 275 ppm. Furthermore, the concentration of carbon dioxide and nitrogen monoxide was also monitored. Finally, in 1967 the California legislation was implemented by the federal government in the Clean Air Act.

In France on August 2, 1961, a framework act was passed which for the first time considered the phenomenon of air pollution as a whole, establishing standards for motor vehicles and domestic heating. The French legislation (updated with the Decree of November 12, 1963) required the control of exhaust emissions for all motor vehicles and also specified the measuring instrument: a photoelectric cell opacimeter approved by the Ministry of Public Works and Transportation. In addition, the measure precisely stated the level of opacity of exhaust gases for each type of vehicle, which was not to be exceeded in any way. A further provision in July 1964 required the monitoring of gases emitted by engine sumps.

In England the Road Traffic Act of 1964 required the upgrading of manufacturing techniques to prevent the emission of visible smoke or gases. It also prohibited the use of devices capable of increasing the fuel flow to engine while in motion (feelers and carburetors not originally fitted by the manufacturers) and decreed that the use of a motor vehicle emitting an excess of gases or fumes was a punishable infringement. The Road Traffic Act, however, did not regulate in any way the emissions of invisible substances that were unanimously regarded by experts as the most dangerous and responsible for the increase in air pollution. In 1964 Belgium and West Germany began to pass strict regulations which declared that all pollutants, solid and liquid, could alter the natural composition of the atmosphere.

The Italian Law No. 615 was passed at a time, the second half of the 1960s, when Italy was slowly developing a growing interest in pollution; the issue received increasingly more space in car magazines. In February 1969 the problem of pollution caused by motor vehicles was discussed during the work of a fact-finding inquiry on the Italian motor vehicle industry promoted by the Twelfth Standing Commission of the Chamber of Deputies.[10] During the hearing, Giovanni Agnelli and Dante Giacosa (the person in charge of industrial planning at Fiat) asserted merely that Fiat was taking "a lot of resources to solve two fundamental problems of air pollution and safety." They also claimed that the internal combustion engine could undergo some improvements, but it would still enjoy "a long life, perhaps very long, for technical and economic reasons."[11] The chairman of Alfa Romeo, Giuseppe Luraghi, was skeptical about the effectiveness of antipollution measures, and he belittled the legislation adopted in the United States. Luraghi, like Giacosa, remained convinced that the "domination" of the internal combustion engine would last for a long time, in spite of the fact that the major car manufacturers had been studying for some time two new engine types (rotary and turbine), and stressed that competition would inexorably limit the development of research.[12]

A representative of the Italian Automobile Union (Sindacato Italiano dell'Automobile) denied manufacturers' claims and—denouncing the backwardness of Italian industry in the study and development of an alternative to the internal combustion engine—said that "the advent of a transformation of the car" could be further delayed only by the "big interests consolidated around the traditional techniques and economies linked to it."[13]

In March 1969 a conference on the pollution caused by cars was held in Milan, promoted by the environmental organization Italia Nostra.[14] The meeting in Milan was characterized by harsh debate. In presenting the published proceedings, Carlo Ripa di Meana[15] highlighted how the issue of pollution from cars was a taboo subject in Italy because of the backwardness of industry studies and for "possible crises in employment and competitiveness of the car industry"[16] following the eventual adoption of mandatory anti-emission devices.

Despite the resistance of the automobile industry, there was a strong debate about motor vehicle pollution. During 1970 the clash between environmental experts and the car industry continued. The incompatibility of their positions clearly emerged out of two important conferences: the first, promoted by the Italian branch of British Petroleum, took place in Milan on April 18, 1970; the second, organized by National Hydrocarbon Corporation (ENI) took place in Rome on June 18 and 19, 1970. The Milanese meeting emphasized opposition to the car; the car industry feared that legislative intervention would excessively increase production costs, as well as discourage the purchase of cars.[17]

The conclusions of the Rome conference were diametrically opposite, since the majority of presentations were designed to demonstrate how pollution would cause harmful effects on health and air quality.[18] It also emerged that the costs of clean air were not prohibitive. One study calculated, in fact, that it was possible to equip all cars with catalytic converters at a per-unit cost of about 37,000 Italian liras, and estimated for the period 1971–1981 an annual average investment of between 19 and 40 billion liras. The same car manufacturers postulated an intervention in three phases: in the first triennium through the adoption of Clean Air Package devices, in the second of systems called Man-Air-Ox, and in the final triennium of catalytic converters. This meant an increase in spending for the period 1970–1985 for the purchase of motor vehicles of approximately 2,000 billion liras and an additional prime cost equal to 2,200–2,400 billion.[19] However, neither the suggestions put forward at the Rome conference nor those of the car manufacturers were implemented. Despite this, and apparently alarmed by a new increase in the cost for car owners, the car magazines took an ambivalent attitude: on the one hand they strongly denounced the existence of the problem; on the other, however, they expressed many doubts about the actual need to adopt legislative measures aimed at limiting motor vehicle pollution. The monthly publication of the Commission for Industrial Motoring of the Italian Automobile Club (ACI), for example, published the General Motors shareholders' meeting report, in which it was alleged that the major causes of air pollution were attributable to industrial fumes.[20]

During 1970 the ACI's weekly magazine wavered between articles with encouraging contents and some polemical pieces with particularly alarming tones. Two editorials signed by Filippo Carpi de Resmini,[21] the president of the Automobile Club, had a great impact. He believed that the issue of pollution could be seriously tackled only through a change in the composition of gasoline to remove lead additives to make possible the installation of catalytic converters on vehicles. De Resmini, however, was not confident of the industry's ability to implement agreements to quickly market these innovations. He also complained that Italian car engines were among the dirtiest in the world, as some models did not follow the standard of eligibility.

The problem raised by de Resmini stemmed from peculiarities in the development of the car in Italy: one of these was the obsolescence of the automobiles in circulation, which chiefly remained (until the 1990s) a constant characteristic of Italian motorization. In Italy the transportation process has always been characterized by its poverty. Whereas in European countries (not to mention the United States) with a high car density, medium- to medium-large-capacity automobiles were predominant, in Italy the majority of drivers owned low-powered cars (up to 1,000 cc). Furthermore, once bought, Italian car owners—whose mean income remained lower than those of other EC countries—were not at all keen to replace them with new ones they could not afford. The opposition to anti-pollution measures was chiefly economic: Italian car owners feared that the adoption of anti-pollution technologies meant higher expenses in their use and upkeep, and first of all that this would require them to replace their cars more frequently.[22]

FROM THE SECOND ANTISMOG ACT TO THE STRUGGLE AGAINST UNLEADED FUEL, 1971–1989

Despite the vibrant tones of the internal debate and the spreading alarm about the unstoppable decay of the urban environment produced by the exponential growth in traffic and air pollution, resistance against legislative measures aimed at lessening cars' environmental impact continued to be considerable for at least two reasons. The first concerned the predominant role of the automobile industry in the Italian economy; until the restructuring of firms which occurred during the first half of the 1980s, the car industry directly employed around 160,000 people (5.7% of the total of manufacturing industries), while indirectly it ensured jobs for two million workers.[23] Public opinion was therefore staunchly against the introduction of any automotive emission regulations because of possible high costs and thus a decrease in car sales, which would generate a crisis for the entire Italian system of domestic production.[24] The second motive is strictly connected with the public's perception of the car as a status symbol. The commercial fortune of the car can be attributed—not only in Italy—to the complex symbolic connotations which it carries. Since its first appearance the car has become the emblem of modernity and the achievement of prosperity. In newspaper and magazine articles, for example, the car was presented as a means for a better life, and it assumed the role of a healing device capable of solving all of life's problems and enabling its owner to become a more modern and even a more highly educated person. The car was presented as the prime symbol of freedom, not only because it enabled its owner to reach any place without public transportation's timetable restrictions, without wasting time, without the effort of carrying luggage by hand, but most of all because the very ownership of a vehicle made one feel like a citizen and not like a subject and conferred the awareness of owning a tool which enabled one to conquer time and space. The car was therefore at the same time cause, effect, and a symbol of social progress, financial prosperity, and freedom. Precisely at a time when, under pressure of the up-and-coming environmental movement, in Europe and the United States the earliest legislation attempted to reduce the environmental impact of motor vehicles through the control of emissions, the car was establishing itself as an emblematic object (object-sign, as Roland Barthes put it),[25] even among the lower classes, as a sort of modern fetish through which one asserted oneself as having achieved prosperity and a new social status.[26]

In this context the European Community gave the final push toward approval of automotive emissions legislation with a guideline (No. 70/220/EEC, March 20, 1970) recommending the unification of member states' laws against such air pollution. The European guideline had been conceived after long and laborious discussions about the abolition of technical standards for motor vehicles by the World Forum for Harmonization of Vehicle Regulations (known as Working Party 29 of the United Nations Economic Commission for Europe, instituted in Geneva in 1952) and by the Working Party Three, instituted within the General Management of Industrial Affairs of the European Economic Community.[27]

To put Directive No. 70/220 into effect, the Italian Parliament passed a law (No. 437, June 3, 1971) mandating the inclusion of devices limiting pollutants among the features necessary to obtain type-approval of vehicles. The law also decreed that tests regarding type-approval had to take place in the presence of officials from the Ministry for Health and the Italian National Health Institute.[28] But Law No. 437 contained provisions designed only to limit the emission of carbon monoxide while ignoring other pollutants such as lead oxides, tetraethyl lead, and nitric oxide present in diesel fuel (diesel vehicles were excluded from the bill).

Meanwhile, between 1966 and 1971, polluting emissions from motor vehicles had increased overall by 46.3%: 6.7% for fine particulate matter, 6.5% for sulfur dioxide, 7.2% for nitrogen oxides, 7.6% for hydrocarbons, 8% for carbon monoxide, and 8% for lead. The region most polluted by sulfur dioxide was Lombardy, followed by Liguria, Veneto, Piedmont, and Emilia Romagna.[29] The conclusions of the *First Report on the Environmental Situation of the Country* (*Prima relazione sulla situazione ambientale del paese*)—drawn by Tecneco, in collaboration with the Ministry of Scientific and Tecnological Research and "under the auspices" of the Prime Minister's Office, highlighted how the approved laws (Nos. 615/1966 and 437/1971) did not offer the possibility of any real control, not even "on the occasion of periodic overhauling." There was in fact "no clear obligation" to owners for the adequate maintenance and upgrading of vehicles, and moreover, no penalty was contemplated "for people driving vehicles and not complying with the requirements established by law." Finally, both Laws 615 and 437 completely ignored motorcycles and established no limits on pollutants other than carbon monoxide and hydrocarbons.[30]

The ACI Stresa Conference took note of the environmental problems of this situation, devoting their thirtieth meeting (September 27–30, 1973) to the subject "The Automobile in Society." The papers presented there focused on analyzing the known data about motor vehicle pollution; Carlo Pollone, deputy director of Fiat's Technological Research Unit, had the task of summarizing technological and legislative developments. Pollone admitted the delay of European car manufacturers in the study of low-emission systems, but according to data presented by American car manufacturers, he stated that installing antipollution devices would have led to a substantial increase in costs for both producers and customers because of higher maintenance costs.

Furthermore, low-emission systems involved major problems for auto manufacturers, such as the high temperatures generated by catalytic converters and the determination of allowable tolerances on various components. In short, Fiat and other car manufacturers were not enthusiastic about the idea of having to equip their cars with new technologies which had already been successfully tested in the United States: Chrysler was the first to develop a modified engine, called the Cleaner Air Package, which consisted of a carburetor calibration that used particularly poor mixtures capable of cutting down carbon dioxide and hydrocarbon emissions. The United States had also developed fuel injection systems, additional devices to control emissions, and appliances to limit the escape of fumes from gas tanks. In Italy the only device that had been designed and applied was the gas recirculation (blow-by) valve. For many years the car industry had already studied some alternatives to the internal combustion engine, especially the "turbocharged internal combustion engines, or Rankine cycle, Stirling engines, electric propulsion, diesel engines with plungers."[31]

Figure 7.1. A famous poster advertising a super-quality gasoline.

Pollone, anxious to demonstrate how the advantage of pollution reduction was counterbalanced by its economic disadvantage, failed to say that the pollution problem in Italy was more serious than in other countries because the Italian fleet was composed mainly of economy cars. These were far more polluting than more powerful cars because the high compression ratio of the engine caused imperfect combustion, the result being a remarkable concentration of pollutants such as nitrogen oxides and unburned hydrocarbons. Pollone could not avoid noting the delay in EC legislation compared to the United States and recognizing that the adoption of the Clean Air Act and the establishment of the Environmental Protection Agency (EPA) had stimulated "an enormous expansion of research and development programs."[32] American laws stipulated minimum air quality standards, provided for the adoption of any necessary measures to meet the approved standards, and required that in 1975 emissions of carbon monoxide and hydrocarbons were to be lowered by 90%, while motor vehicles produced in 1976 would have to emit 90% fewer nitrogen oxides than allowed in 1971.

European legislation was far more permissive: it required a reduction in carbon monoxide and hydrocarbon emissions of only 50% and 35%, respectively, than in vehicles not fitted with low-emission systems. Furthermore, it called for recirculation of exhaust gases and laid down a safety limit of 4.5% carbon monoxide emitted by idling engines, which many experts judged too high and extremely harmful to health. While European legislation imposed a further reduction in carbon monoxide and hydrocarbons present in car exhaust starting from 1974–1975, it did not establish any limits for nitrogen oxides. Engineer Pollone, not without surprise and disappointment, declared that Sweden and Australia had decided "for reasons not clearly understood"[33] to align with US standards. The reasons were not so mysterious, but lay in concerns about health hazards; in the United States some new studies, with an increasing wealth of data, showed the hazards of car exhaust.[34]

Since the late 1970s, car magazines—whose attitude was constantly torn between the demand for antipollution measures and protecting the interests of car manufacturers—reserved an ever-larger amount of space to the growing problem of motor vehicle pollution. In early 1977 *Quattroruote* published two articles devoted to the hydrogen engine in which it recalled how its first patent, applied for by the American A. F. Burstall, dated back to 1927 and that in Italy, the engineer Massimiliano Longo had successfully tested a hydrogen engine in an Alfa Romeo 1300 Giulia. In December 1977 *Quattroruote* reported on work on the electric engine by Fiat, Ansaldo, Enel, Garelli, Magneti Marelli, and Piaggio Zagato. The result was a series of prototypes such as Zele Zagato (with a range of 60 km), Fiat XI 23 (excellent performance and a range of 70 km), the Fiat-Enel van (range of 50–80 km and operating costs equal to 3,000 Italian liras per 100 km), and the Taxi PGE (range of 100 km at an average speed of 50 km/hr). In July 1979 *Quattroruote* gave great prominence to the Fiat 131 model powered by a hybrid electric-gas engine presented by the company at the Detroit Motor Show.[35]

In 1983 the controversy about the harmful effects of leaded gasoline reached Italy. The issue was raised in 1981 by British consumer associations with the dual aim of denouncing the high toxicity of leaded gas and obtaining the revocation of an European Economic Community (EEC) rule which forbade member states from producing unleaded fuel. In April 1983, after a study had found that cars were responsible for 70% of the lead emitted into the atmosphere, the British government passed a law making it compulsory to use unleaded gasoline for all cars produced from 1986 on. Furthermore, the law set a 1995 date for the withdrawal of leaded gasoline from the market.

The British government's decision was shared by all other European countries except for Italy and France. Germany also imposed the adoption of catalytic converters (to limit the emission of five substances: nitrogen oxides, carbon monoxide, unburned hydrocarbons, sulfur dioxide, and aldehydes) and required lead-free fuels in all cars produced beginning in January 1986. The measures were motivated by the need to drastically reduce the chemicals responsible for the acid rain (nitrogen oxides, unburnt hydrocarbons, sulfur dioxide) that was severely damaging German forests. The Italian government had not followed other member states in order to support Italian car factories, who requested the use of unleaded gasoline with a high octane rating (around 96), whereas Europe diminished the amount to 92.

The solution to the car pollution problem still seemed very far away because the production of unleaded gasoline sparked a bitter controversy between those who wanted to encourage its use in cars not fitted with catalytic converters and those who sought to restrict the use of unleaded fuel to cars with them. The crux of the matter was the benzene and other aromatic hydrocarbons (highly carcinogenic substances) found in unleaded gas which served to maintain the desired high octane. Using these fuels to power cars not equipped with catalytic converters therefore meant filling the air with highly toxic and dangerous substances, in addition to lead. Car manufacturers, along with refiners and even politicians, were aware of this trade-off.[36] In fact, in April 1989 the findings of the Commission of Inquiry on Fuels were released, which proposed reducing benzene from a maximum of 5% in volume to 3% in weight and considered undesirable the use of unleaded gas in cars not fitted with catalytic converters as this would have resulted in increased pollutants (mainly aldehydes and aromatic hydrocarbons).

Consequently Minister for the Environment Giorgio Ruffolo suggested the government put a ban on the sale of unleaded fuel for those cars, adopt a measure establishing a maximum level for aromatic hydrocarbons, reduce the price of unleaded gas, and allocate sales incentives for the purchase of catalytic converter–fitted vehicles. Some bills were also introduced to limit benzene to 3% in weight and aromatic hydrocarbons to 30%, in observance of the European directive that unleaded gasoline should not represent an additional source of pollution.

Following a bitter campaign promoted by the Italian Oil Union (Unione Petrolifera) on July 12, 1989, Minister Ruffolo signed a letter of intent for the reduction of aromatic hydrocarbons which set the maximum allowable benzene at a higher level, 4.5% in weight, and that of other hydrocarbons at values set for unleaded gas in other EEC countries. It was an insulting measure, when in Japan and the United States the maximum level of benzene was 1.5%. The rules were too lax and unable to produce a real reduction in the amount of aromatic hydrocarbons in the air.

In October 1989 a response to such an ineffectual decision by the government came from *Quattroruote*, which proposed its own antipollution therapy focused on a clean engine without a catalytic converter. The proposal was a compromise between the positions of environmentalists and those of Italian car manufacturers, who had always been very lukewarm toward low-emission technologies. Meanwhile, with inconclusive ideas and farcical measures, the pollution problem grew more serious.

FROM THE 1990S TO THE EARLY TWENTY-FIRST CENTURY

In 1990, Italy emitted 430 million tons of carbon dioxide into the atmosphere (96 million from transportation), 1.938 million of nitrogen oxides (1.141 million from transportation, including 945 million just from motor vehicles), 2.192 million of non-methane volatile organic compounds (1,121 million from transportation, including 954 million from vehicles), and 1,653 million of sulfur oxides (124 million from transportation, 102 million from vehicles). Motor vehicles therefore represented a significant source of pollution: in 1991 they produced 643.3 million tons of carbon dioxide, with 82.3 million tons emitted by aircraft, 20.7 million by ships, and 8.7 million by trains.[37] Air quality was affected in almost all medium to large urban conurbations, with Milan as the worst affected, then Rome, Naples, Turin, Bologna, and Florence. In Milan and in thirty-four other municipalities of the Milan hinterland, values of carbon monoxide exceeded 20 mg (the alarm threshold was 10 mg), and those of nitrogen dioxide permanently surpassed the 400 μg limit set by the WHO.[38]

Despite the large number of people opposed to antipollution measures, in 1994 a ministerial decree (No. 114, April 15, 1994) was approved to establish an urban system for monitoring carbon monoxide, sulfur dioxide, nitrogen dioxide, ozone, and fine particulate matter. Once these levels were exceeded for at least three consecutive days, municipalities had to adopt mandatory antipollution measures.

Local authorities tried to decrease pollution levels with some buffer solutions, such as having traffic on alternate days (*targhe alterne*)[39] and the restriction of traffic during some times of the day and on Sundays ("Sundays on foot"). Communes introduced a series of imaginative proposals to reduce traffic, such as sales incentives for electric vehicles, shared taxis, and entrance fees for historic centers (eco-pass). These measures received angry protests from professional and trade associations (especially from truck and taxi drivers and merchants) and from citizens not allowed to use their cars.

Since the second half of the 1990s there has been a considerable increase in the sales of medium-high- to high-powered cars, above 35% in the northern regions. This clearly indicates that consumers' choices are especially inclined toward vehicles capable of undertaking daily long journeys and that provide more comfort and safety. From an environmental viewpoint, however, the proliferation of large and powerful cars has caused more parking difficulties, increased road congestion, higher fuel consumption, and more pollution. These vehicles (SUVs included) are used in cities for going only short distances with just the driver on board, resulting in a higher contribution to pollution in urban areas and more road congestion. The proliferation of such vehicles has undermined environmental improvements due to the response to the more recent European recommendations (Euro 3 and Euro 4).[40] It is clear that on the basis of equal power and better fuel consumption, the new-generation vehicle has a smaller environmental impact than that of older vehicles; however, this positive factor has been partly negated by the proliferation of powerful cars and by those fueled by diesel.

Notwithstanding Italy's application of European recommendations in matters of air pollution with Ministerial Decree No. 60 of April 2, 2002—setting much more severe limits on the primary pollutants—almost all major cities continue to disregard the limits imposed by these guidelines. Nitrogen dioxide rises well above the established limits in almost all major Italian cities. European legislation has established an ozone limit of 120 $\mu g/m^3$, not to be exceeded for more than twenty-five days in a year; this limit is observed in only four Italian cities. Fine particulate matter (PM_{10}) remains the pollutant presenting the gravest problem. Notwithstanding the growing number of measures of a nonstructural nature aimed at reducing it (e.g., limitations on traffic, alternate number plates), such measures are ineffective as no city is able to remain within the limits imposed by legislation (50 $\mu g/m^3$, not to be exceeded for more than thirty-five days in a year).[41]

* * *

Despite the constant updates to legislation and the introduction of new measures aimed at containing pollution within an acceptable threshold, the environmental crisis caused by vehicular traffic is still far from being resolved. While replacing leaded with unleaded gasoline, improving diesel fuels, and using catalytic converters has decreased the concentrations of lead in the air, carbon oxides, nitrogen dioxide, sulfur dioxide, and other pollutants such as benzene and PM_{10} remain well above hazardous levels.

To radically reduce pollution, there is only one way forward: the replacement of internal combustion engines with new engines powered by electricity or hydrogen. In the immediate future, however, it is unlikely that these technologies will be adopted on current models: electric cars continue to have high costs and a limited range, while those that do not produce hydrogen gas on board (hydrogen is placed in a refrigerated tank at –253°C) are limited by the lack of a distribution network, which is still far from being developed. Hydrogen engines with reformer technology (a sort of chemical plant that draws fuel from methanol or gasoline) have unsustainable costs that prevent any kind of mass production.[42]

The slow dissemination of hybrid models and alternative fuels—which include the little-used reformulated gasolines containing low amounts of aromatic hydrocarbons, replaced by plant-based additives, and white diesel (a mixture of water, oil, and additives)—suggests that the clean car is destined to remain utopian for many years to come.

PART III

POLITICS

Since the crisis of 1992–1994, when a judicial investigation called Mani Pulite (Clean Hands) brought down the party system established in the aftermath of the fall of the Fascist regime, grassroots movements have developed in Italy which are opposed to the building of infrastructures. They cite two reasons for their action: one is their fight against corruption (in their view, by obstructing the implementation of public works they consequently get rid of shady deals and kickbacks between contractors and the representatives of state institutions and political parties); the other is protecting the environment. Politically, these groups gravitate toward both the galaxy of tiny parties to the left of the Democrats and anti-system organizations such as the 5 Stars Movement, but what they have in common is a nebulous anticapitalism and above all a rejection of industrial society.[1] The two main pillars of these grassroots groups' ideology are the defense of common goods (deriving from participatory experiences against locally undesired uses of the land) and "de-growth." They maintain that common goods (atmosphere, water, forests, biodiversity, physical and digital infrastructures, and local commons such as hydrographic basins and urban ecosystems) are not monetizable. In this vision the dominant economic paradigm is countered by an ecological and communal view of the world; resistance to locally undesired uses of land may therefore be considered as one of the tools for concretizing opposition. *De-growth* is understood by

Serge Latouche as "a slogan of political provocation whose purpose is chiefly to make us find once more a sense of limit. . . . [I]t is clear that de-growthers want growth in the quality of life, of the air, water and a multitude of things that growth for growth's sake has destroyed. . . . In other words, right from the start we're not talking about an economic plan, though it is another economy, but about a social plan that implies an exit from the economy as a state of things and imperialist argument. . . . Nonetheless, the de-growth society is neither a return to the past (impossible), nor an alignment with capitalism, but rather a superseding (in good order if possible) of modernity. De-growth is necessarily against capitalism."[2] These two pillars make an actual chimera of the movements' ideology: half nondeformable ideology, half utopian narration of a hypothetical world.

Other shared matrices are identifiable in certain ideologies of radical environmentalism, such as primitivism (which advocates a return to a pre-agrarian social model), deep ecology (whose objective is to subvert the current against-nature civilization), vegetarianism (especially its most radical offshoots such as veganism and fruitarianism), which have all become actually mainstream, an alternative to the system of cultural and scientific values imposed between the seventeenth century and the 1970s.

Appropriation of these theories by grassroots groups (but often trivialized and reduced to cool fashions) contributed decisively to a decline of the collective environmentalism that flourished in the 1980s, with vertical national and international organizations where the central structure and branches, run by full-time militants, are flanked by the wide grassroots participation of volunteers, which helped to make the organizational structures less formal. The rapid transformation of environmentalism in a radical direction was also encouraged by the literature (environmental historiography included), which began to interpret environmental transformations through the rigidly eco-centric perspective that was becoming increasingly widespread within political ecology.

Grafted onto this picture was the traditional Italian mistrust of the intellectual professions, which have always been seen as the product of an elite whose interests contrast sharply with those of the people. As the historian Ernesto Galli della Loggia has quite aptly written, politics in republican Italy—represented by "tight national elites, educated and hand-picked"—succeeded in organizing a society pervaded by "impetuous but voracious animal spirits and as a whole boorish and mannerless, home to powerful criminal organizations, a society localistic in the main and animated by strong acquisitive instincts." This situation was overturned by the Clean Hands investigation which definitively assigned primacy to society, thanks to "the contemporary ascendency throughout the western world of an antistatist ideology," and "to the radical delegitimization of politics, of its tools and apparatuses."[3]

Setting out from these presuppositions, this third part comprises three chapters that analyze certain sociopolitical dynamics linked to public decision-making processes and to governance of the region. Chapter 8 deals with the relationship established among experts, politics, and institutions, in which historical-environmental literature has only sporadically taken interest in the question. The exiguity of studies on the relationship between experts and politics might be put into relation with the epistemological difficulties which history of the environment encounters in declaring its status as a new discipline deriving from the hybridization of techno-scientific and sociohistorical knowledge, as well as with the current preeminence of an ecocentric-militant approach. The latter contributes to distancing those scholars who are interested in studying environmental problems but are not inclined to produce a historiography that identifies itself as a sort of academic rib of political environmentalism whose parabola, from the second half of the 1990s, has been increasingly portrayed in an antagonistic sense. The chapter, centered on the scientific knowledge inherent to environmentalism and on town planning, seeks to provide a first reflection, far from any claim to completeness or exhaustiveness, on the relationship between experts and political decisions in contemporary Italy. Chapters 9 and 10 are dedicated to the transformation of the environmental movement, characterized by two phenomena: the progressive institutionalization of environmental organizations and the appearance of numerous protest movements at the local level which, gradually, have taken on an increasingly particularistic and localistic character. These movements' origins are often referred to by the acronyms LULU (Locally Unwanted Land Uses, neutral from the evaluative viewpoint) and NIMBY (Not in My Back Yard, whose accepted meaning has a negative connotation). In the climate of permanent electoral campaign, a feature of the Italian political debate since the 1990s, the birth of these movements has complicated the situation by helping to radicalize environmentalism and by interpreting environmental transformations in a rigidly ecocentric perspective. In this way the dynamics at the core of environmental problems are increasingly filtered through the lens of an ideology that molds them to the public use it seeks to promote.[4]

CHAPTER 8

ENVIRONMENTAL KNOWLEDGE, URBAN PLANNING, AND POLITICAL DECISIONS IN ITALY FROM 1850 TO THE PRESENT

Historical-environmental literature has only sporadically dealt with the relationship among experts, politicians, and institutions, although questions concerning the environment and the land are of a complex technical nature.[1] This is surprising inasmuch as history of the environment is not a humanistic discipline in the traditional sense of the term, being largely practiced by scholars from the natural sciences, physical sciences, and techno-scientific disciplines. This chapter centers on the scientific knowledge inherent to environmentalism[2] and on town planning and seeks to supply a first reflection, far from any claim to completeness or exhaustiveness, on the relationship between experts and political decisions in contemporary Italy.

THE DECADES OF THE TECHNOCRACIES, 1870–1942

Over the course of the nineteenth century the progressive spread of ideas critical of creationism and mechanistic materialism set the conditions for the birth of ecological thought and movements to protect the natural environment. In 1866 the German zoologist Ernst Heinrich Haeckel (1834–1919) began using the word *ecology* to indicate the relationships between living organisms and the environment, and in 1893 the president of the British Association for the Promotion of Science raised "ecology to the rank of one of the three great parts of biology, on the same level as physiology and morphology."[3]

In the second half of the nineteenth century, primarily in the United States, two philosophical-cultural visions of the relationship between man and nature were developing: the first, preservationism, held that it was indispensable to preserve those portions of nature that had remained inviolate and thus stem the destruction of the land by industrial development and urbanization; the second, conservationism, subordinated protection of nature to an economistic management of natural resources with the objective of rationalizing human interventions in the ecosystem to maintain its biological equilibriums.[4] The progressive spreading of these ideas led to the establishment of several associations for the defense of nature and to approval of the first interventions to protect the environment. These organizations took their first steps in Great Britain, partially because of the rapid industrialization which had had significant consequences for the country's natural and urban environments.[5] The progressive spread of conservationist policies resulted in two international meetings, both held in Paris: the first (1883) on protecting seals in the Bering Sea and the second (1895) on protecting species of birds useful to agriculture. In the course of the 1910s and 1920s the fortune of conservationist ideas and the progressive success of ecological science paved the way to certain initiatives of international breadth: in 1913 a Permanent Consultative Commission for the Protection of Nature was established in Berne, proposed by the Swiss naturalist Paul Sarasin at the Eighth Congress of Zoology in Graz (1910); in 1923 the First International Congress for the Protection of Nature was held in Paris; and in 1928 an International Office for the Protection of Nature was opened in Brussels. In London in 1933 Great Britain, France, Belgium, Italy, Spain, Portugal, South Africa, Egypt, and Sudan signed a Convention Relative to the Preservation of Fauna and Flora in their Natural State through the control of trade in hunting trophies, prohibition of certain methods of capturing animals, and the creation of protected areas. In 1940 twenty-two countries in the Americas, including the United States, Mexico, Argentina, and Brazil, signed in Washington a Convention on Nature Protection and Wild Life Preservation in the Western Hemisphere, with a view to saving flora and fauna of the American continent from extinction.

In this context a nature protection movement also began to develop in Italy, spearheaded by the Associazione Nazionale pro Montibus et Silvis (established 1898) and the Italian Touring Club (founded 1894 in Milan). Both associations contributed to spreading a holistic and aesthetic vision of nature which, together with an idealized representation of the lifestyles typical of rural societies, formed the ideological foundations of Italian conservationism. Their main goal was the protection of the environment (understood prevalently as landscape) because it was a possession of inestimable value inherited from the past.[6] Between the end of the nineteenth century and the 1930s the question of preserving nature lay at the center of a composite network of relationships that united associations, intellectuals, functionaries, and parliamentarians to such an extent as to constitute, as Luigi Piccioni has written, "an actual protectionist lobby" present in ministries, in Parliament, and within the purview of political parties. It was an actual technocracy, set in a cosmopolitan dimension, convinced that "conciliation between the exigencies of progress and 'beauty' and/or of protecting nature [was] possible and proper." Piccioni groups these elites of environmentalist pioneering under three headings: naturalists, artistic-patriotic people dedicated to protecting historical and artistic heritage, and supporters of positive economic exploitation of natural beauties through the promotion of tourism.[7]

In parallel with environmental protectionism, a hygienist movement was coming to the fore, backed by professional figures (e.g., physicians, engineers, architects) who were acquiring a growing social role in the development of urban and industrial society.[8] In the wake of the progressive success of hygiene[9]—understood as a scientific discipline that studies the health of human collectivities—a gradual process of urban renewal began which, between the last two decades of the nineteenth century and the 1940s (the period Martin Melosi has called "the bacteriological revolution"),[10] led to the building of the "healthy city" by means of what Gabriella Corona and Simone Neri Serneri have described as "a radical reorganization of ways to use human resources, setting out from the reconstruction of water adduction and emission circuits and proceeding right down to the covering of the soil, to waste management and the siting of the 'unhealthy' production plant."[11] In spite of the tensions produced by the processes of incorporating nature into urban-industrial society,[12] the interactions among technocracies, economic players, national politicians, and public administrators gave rise to certain, not inconsiderable, creations: health legislation aimed at protecting waterways, soil, and the atmosphere against improper uses not compatible with the collective interest;[13] a corpus of laws for protecting cultural heritage;[14] the establishment of national parks;[15] and the town planning law of 1942.[16] Although historiography does not fail to point out the lacunae and contradictions inherent in these provisions, it seems appropriate to underscore the technocracies' ability not only to insert cosmopolitan requests into the Italian public debate but to actually translate them into concrete actions—in the absence of even minimum popular consensus—thanks to efficient lobbying.[17] These results were not marginal and were achieved in a country characterized by a late process of national unity and by complex cultural and social disparities, as well as by profound regional dissimilarities.

ACCELERATED DEVELOPMENT AND THE IRRELEVANCE OF EXPERTS, 1950–1980

In the years of the economic miracle the paradigm of modernization[18] and the rhetoric of modernization marked the way to development; the imperatives were to reconstruct the country, modernizing its infrastructures and lifestyles to bring Italy into the assembly of industrialized and developed nations. In this context there was a collective repression of the environmental damages, fed by awareness that the benefits of the development model—in the eyes of the great majority of citizens and of political and economic policy makers—far outweighed the disadvantages. The picture began to change, very slowly, only around the mid-1970s when the environmental crisis became evident—revealed in the second half of the decade by the water resources crisis as well as by the pollution of water, air, and soil—and the necessary remedy could no longer be put off.

In the decades of accelerated development the environmental movement—structured around several national organizations (Italia Nostra, Movimento Italiano Protezione della Natura/Pro Natura, Lega Italiana Protezione Uccelli (LIPU), World Wildlife Fund [WWF])[19] that remained influenced by the internationally predominant conservationist approach[20]—found itself operating in a social and political context fundamentally indifferent to environmental problems. As Mario Diani has observed, there was total absence of a "cultural orientation favorable to environmental defense that [had] any mass consistency," and within national associations, there was "an almost total lack of basic action by citizens."[21] From the first half of the 1970s—in the wake of the nascent debate on the stationary society and the finiteness of natural resources[22]—the word *ecology* began to appear in public debate, and in parallel with the publication of ecological research,[23] the conceptual foundations of environmentalism began to be transformed by the integration of scientific ecology into political and social theory. Political ecology—whose birthdate is generally agreed as being the publication of Rachel Carson's *Silent Spring* and Barry Commoner's *The Closing Circle*—considered environmental problems closely correlated with industrial and urban development and saw, in the nascent environmental movement, a social force capable of influencing the political debate to press demands for going beyond the model of capitalistic development. The Amici della Terra was founded in 1977 and, together with the Antihunting League (LAC) and the Antivivisection League (LAV), represented an environmental pressure group associated with the Radical Party. The League for the Environment (subsequently Legambiente) was set up in 1980 within the Italian Recreational and Cultural Association (ARCI) and right from the start showed itself to be a flexible organization, dealing with multiple environmental issues ranging from town planning to conservation of natural resources, pollution, and the promotion of alternative lifestyles. The advent of political ecology organizations spurred the conservationist associations to include such themes as criticism of the industrialist development model and the crisis of natural and energy resources.[24]

As for the management of cities and the land, sanitary engineers had passed the baton to the town planners, gathered within the National Town Planning Institute (INU), "the happy place where technical wisdom and civic passion had united."[25] Founded in 1930 on the occasion of the Twelfth International Congress of the Féderation Internationale pour l'Habitation, l'Urbanisme et l'Aménagement des Territoires, the INU was working during the years of accelerated development to overcome the limits of the law of 1942 through three basic principles: the need to coordinate town planning with national, regional, and area-specific economic programming; the implementation of town planning at area-specific and regional levels; and the creation of public building laws to arrest the wildcat growth of cities and curb land speculation.[26] During the golden age of industrial capitalism, the environmental experts and professional groups, notwithstanding their close interaction with economic players and political parties, did not succeed in achieving significant results. The environmental laws approved during this time were Nos. 615/1966 and 437/1971 on atmospheric pollution and No. 319/1976 on water protection. The first provisions were bitterly criticized by scientists as being merely of a general nature, little more than statements of intent. Law No. 319 was progressively voided through a series of deferments, exceptions, and modifications.[27]

In an earlier work I employed the expression *paper environmentalism* to define the environmental policies implemented between 1950 and 1980 and to underscore how, as much at a national as at local level, these were little more than a rhetorical expedient to hush up the functionaries and offices delegated to study ecological problems, as well as those minority sectors of science and society that were beginning to denounce the ecological crisis. The environmental rhetoric of the institutions, in a word, assumed the function of a communications maquillage aimed at public opinion, whereas formal actions were still firmly centered on the problems of employment and production, to which ecological matters were subordinated.[28]

The reform projects for town planning were never approved: the Fiorentino Sullo was left stranded by a construction industry offensive and the Christian Democrats' decided to withdraw support from the Ministry of Public Works' proposal, while the draft law presented by Giovanni Pieraccini was dropped by the Moro governments.[29] The municipal town plans—approved with difficulty and soon stripped of meaning—did not succeed in stemming the disfigurement of urban areas. The result, as Salzano has written, was that "the ideal and political force" was cut off, and the INU fell back "on debates increasingly closer to the academy, increasingly mainstream on the technical side, increasingly polluted by evasion."[30]

To grasp the irrelevance of experts in environmental and town planning during the golden age—but also the complex interweaving of national economic interests, international pressures, and the need to give a concrete response to social changes that occurred more rapidly than ever—we might make reference, in the words of Roland Barthes,[31] to a *sign-object* of modernity and modernization: the automobile. The potential problems of mass motorization—pollution of the atmosphere, the human costs of traffic accidents, the transformations of the land brought about by traffic flows, and urban sprawl—had been well known and analyzed since the early decades of the twentieth century. The first studies on polluting emissions were published in the 1920s, and in 1950 the California biochemist Arie Haagen-Smit demonstrated the existence of a direct relationship between vehicle exhaust and the smog enveloping Los Angeles: that photochemical smog was produced by the reaction of hydrocarbon compounds and nitrogen oxides with sunlight.[32] The relationship between accidents and the unsafe car models on the road was denounced in 1965 by Ralph Nader in a book that triggered a heated international debate.[33] A considerable bibliography on the relationship between the development of private car ownership and transformations of the urban form already existed as early as the 1970s.[34] Notwithstanding the numerous studies produced (perfectly in line with the results of the most advanced international research and widely disseminated), neither the central government nor local bodies paid any attention to experts' recommendations, abstaining from making choices capable of upgrading the urban ecosystem to accommodate the transformations imposed by a growing demand for private mobility. During the course of the 1970s the few measures adopted by municipal administrations failed because of their fragmentary and antiscientific nature.

This failure of urban and environmental policies cannot be imputed solely to the break in relations between experts and political players. The latter, in order to mitigate the great social tensions produced by accelerated economic growth, gave priority to the satisfaction of citizens' primary needs (housing first and foremost and the creation of new jobs), but also to the lobbying activities of economic players who acted mainly at the supranational level. As Sigfrido Ramírez Pérez has demonstrated, approval of regulations for the improvement and homogenization of technical standards for protecting the environment, such as energy efficiency and vehicle safety, was slowed down by a decade-long debate which saw the producing countries clashing at the World

Forum for Harmonization of Vehicle Regulations (also known as Working Party 29 of the United Nations Economic Commission for Europe) and at Working Party Three set up at the EC Directorate-General of Industrial Affairs.[35]

THE DECADES OF EUROPEAN SUASION, FROM THE 1980S TO THE PRESENT

During the course of the second half of the 1980s the environmental movement made a great leap forward thanks to the nuclear question. Between the end of the 1970s and the early 1980s the movement had favored a fusing of conservationist, environmental, political, and social elements that translated into the mass mobilization of the first referendum campaign, promoted by Amici della Terra and the Radical Party, for abrogation of the regulations on the location of nuclear power plants. The referendums of November 8 and 9, 1987, ratified the abandoning of nuclear energy and marked a clear success for Italian environmentalism that was reflected in their numerical growth: around the mid-1990s in fact the WWF reached 281,000 members, Legambiente 115,000, and the Italian branch of Friends of the Earth 25,680.[36] In the wake of victory against nuclear power, the Federazione delle Liste Verdi won 2.5% of the Chamber (thirteen seats) and 1.96% of the Senate (one seat), while certain leading figures of environmentalism were elected to Parliament from the Communist Party (Laura Conti, Enzo Tiezzi, Giorgio Nebbia, and Enrico Testa) and the Proletarian Democracy Party (Edoardo Ronchi and Gianni Tamino).

The Italian environmental movement appeared dominated by organizations of a national character distinguished by a strong central structure and extensive regional ramifications. Predominant among these (at least numerically) were the ones with a conservationist approach, first and foremost the WWF whose structure, in Diani's view, was the only one comparable to the environmental lobbies in English-speaking countries. It was followed by the political ecology organizations, the most significant undoubtedly being Legambiente, which over the years had promoted an important series of campaigns aimed at legislative interventions for environmental protection. Then there were militantly active organizations such as Greenpeace Italia (1986) and Mountain Wilderness (1987) whose approach to environmental problems was of a multinational nature and favored demonstrative action rather than lobbying. This green archipelago was completed by associations that worked in sectors contiguous to those of the environmental movement (Fondo per l'Ambiente Italiano [FAI], Touring Club, Club Alpino Italiano) and by right-wing environmental groups.[37] From the mid-1990s the transformation process of the environmental movement was characterized by two phenomena: the progressive institutionalization of their organizations and the appearance of numerous protest movements at the local level, which gradually assumed an increasingly particularistic and localistic character.

A vast range of literature maintains that expansion of the environmental movement coincided with the transformation of associations into "formal, professional and tendentiously centralized" organizations, which is to say run by great bureaucracies with high levels of professionalization.[38] The progressive institutionalization of environmental associations—having become NGOs with a structure very similar to that of a traditional political party—led a growing number of citizens to distance themselves from the formalized environmental movement and seek alternative forms of protest, generally conducted by civic committees.[39]

The 1980s saw the beginning of what Vezio De Lucia has called a "counterreformation" in town planning, meaning setting aside the processes and proposals fine-tuned in previous decades, thereby creating a "sort of oxymoron," which is to say the "privatization of town planning, . . . town planning entrusted to private parties."[40]

Presently the judgments of the protagonists of regional planning leave no space for optimism. In Vezio De Lucia's view, although "it is inappropriate to speak of a town planning crisis . . . there is a crisis in the ability of public powers to respond to distress about the urban condition. The official apparatuses of the State are in crisis, as are almost all the local bodies with regard to public services, starting with transportation. [We have] a crisis that involves cultural organizations, from universities to the National Town Planning Institute."[41]

Giuseppe Campos Venuti, in dialogue with Federico Oliva, acknowledges

> that town planning, from the fifties to the seventies, was considered a theme of great interest in the country. . . . Restricted cultural and university circles paid attention to town planning then, as did the country's economic, political and trade union leaders, and the press with dailies and weeklies for intellectuals; in brief, a narrow circle of people who made decisions about this interest, without taking into account public opinion. . . . That the interest in town planning has unquestionably waned, however, nowadays regards a fairly wide segment of public opinion that is informed about the country's situation by television, whose control is unquestionably in the hands of a narrower and much less pluralistic group of economic and political leaders in comparison with the past, while the influence of the press has been heavily reduced, and the cultural and university milieus have grown in number, but now count for very little in terms of directing public opinion.[42]

In this picture concrete implementations have run aground in the face of governments' inability (including those of the center-left from 1996 to 2001 and from 2006 to 2008) to approve new town planning reform, and in the face of the constitutional modifications introduced by Law No. 3 (October 18, 2001), which permitted approval of regional town planning laws. The result has been the country's government losing its national dimension, instead being fragmented into a warren of often very dissimilar regional approaches. Deregulation has thus increased private influence over the management of building policies.[43]

The parabola of environmental policies has followed a different trajectory, due to two external factors whose importance turned out to determine the definition of concrete realizations. The first is represented by the United Nations which, beginning with the Stockholm Conference on the Human Environment (1972), succeeded in promulgating a global perspective on environmental problems. The United Nations Environment Programme (1972)[44] was able to elicit dialogue among the scientific community, technical experts, and policy makers. Perhaps the best-known outcome concerns the United Nations Framework Convention on Climate Change which in 2015 concluded a universal agreement on the question (COP 21, Paris). The second concerns European Economic Community (EEC) action, of fundamental importance for the broadening of environmental legislation in member states.[45] In Italy too (indeed one could say above all in Italy and the other countries of Mediterranean Europe) the environmental regulations since the 1980s have been progressively defined and updated on the basis of European directives. The importance of community policies is evident in the Italian case: in the absence of the European obligation, in fact, there would have been much greater resistance to approval of an efficacious body for environmental legislation. Without the political persuasion of the EEC—with which Italy had to comply in observance of signed treaties—the pressure groups that had delayed application or weakened environmental laws promulgated in the 1960s and 1970s would have continued to procrastinate the approval of incisive regulations.

However, in spite of EEC suasion, environmental crises continue. For this reason, in a framework characterized by progressive public mistrust of national and local policy, the experts are increasingly involved in environmental conflicts: on the one hand they represent the professional figures who propose infrastructure projects of public interest, while on the other they are engaged by environmental associations and civic committees who have recourse to technical-scientific knowledge in attempts to back up their own theories that are against development. Often the experts engaged by environmental organizations turn out to be more authoritative and credible than those hired by politicians: “the only expertise that takes the floor in a systematic manner and with recourse to technical-scientific arguments is the one that opposes the project, presenting more numerous and detailed assessments, questioning the overall design of the work, the conditions and objectives that justify it.”[46] So paradoxically “the technicalization of a controversy may both cushion and stoke a conflict, close down conversational spaces or contrariwise open them, exposing the political matrix of the same technical options inasmuch as they are, precisely, options and not necessities.”[47]

CHAPTER 9

THE PARABOLA OF ITALIAN ENVIRONMENTALISM

FROM THE CENTRALITY OF ASSOCIATIONISM TO THE SUCCESS OF NIMBY LOCALISM

In the second half of the nineteenth century, and mainly in the United States, two philosophical-cultural visions of the relationship between man and nature were developing which may be summarized under the names of preservationism and conservationism. Preservationism began with the assumption that ecological problems originated from an anthropocentric culture, considering it indispensable to preserve those portions of nature that had remained inviolate and thus stem the sacking of the land brought about by industrial development and urbanization.[1] Conservationism's chief objective was to encourage action aimed at rationalizing human interventions in the ecosystem so as to maintain its biological equilibriums.

"GREEN PIONEERING": THE EMERGENCE OF CONSERVATIONISM

The foremost representative of conservationism was Gifford Pinchot (1865–1946), director of the US Forest Service. In the early decades of the twentieth century Pinchot contributed to the emergence of progressive conservationism, whose ideological foundations may be identified in a rigidly dichotomous representation of the animal kingdom in which there were "good" animals (birds, cervids) and harmful ones (carnivorous predators), and in the economic concept of nature as a complex of resources to be profitably managed.[2]

The 1930s marked a fundamental turning point in the history of the environmental protection movement: the incremental emergence of the idea that safeguarding nature was an ethical imperative (rather than merely an instrument for centralized and production-oriented management of natural resources) strained the credibility that conservationism had enjoyed. Certain essays began to underscore the fallacy of the economic approach of environmental conservation and attempted to go beyond the previously dominant anthropocentric perspective by outlining an ecocentric and holistic conception of nature, understood as a system capable of harmonious self-regulation.[3]

These ideas met with such great fortune that the term *conservation* began to be used to indicate an environmental protection whose objective was the maximum reduction of anthropic pressure on a natural ecosystem in order to maintain its ecological balances.[4]

Such a nature protection movement began to develop in Italy as well, where its cornerstones were the Associazione Nazionale pro Montibus et Silvis (established in 1898 in Rome) and the Italian Touring Club (founded in 1894 in Milan).[5] Both associations contributed to spreading a holistic and aesthetic vision of nature which, together with an idealized representation of the lifestyles typical of rural societies, formed the ideological foundations of Italian conservationism, whose main goal was protection of the environment (understood prevalently as landscape)[6] because it was a possession of inestimable value inherited from the past.

Early in the first decades of the twentieth century the protectionist movement carried out a series of campaigns, such as demanding the creation of national parks in Italy and the preservation of the Ravenna pinewoods and fighting the pollution caused by an aluminum factory that the Montecatini company had built in Val Lagarina (a part of the Adige Valley between the Lessini Mountains and Mount Baldo).[7]

The development of the protectionist movement coincided with approval of the first environmental protection laws, in particular the legislation on national parks begun with a royal decree on December 3, 1922 (Establishment of a "National Park" in the Zone of the Gran Paradiso Group in the Graian Alps, which became Law No. 473, on April 17, 1925), and continued with laws establishing the parks of Abruzzo (Law No. 1511, July 12, 1923), Circeo (Law No. 285, January 25, 1934), and Stelvio (Law No. 740, April 24, 1935). In June 29, 1939, Law No. 1497 on Protection of Natural Beauties protected natural environments of special aesthetic or geological interest and also man-made environments of "uncommon beauty" (e.g., villas, gardens, parks), with reference to the Italian landscape "tradition" (e.g., the organization of urban blocks and the building techniques characteristic of different regions).

"ASSOCIATIVE ENVIRONMENTALISM": THE GOLDEN AGE OF THE ENVIRONMENTAL MOVEMENT?

In the aftermath of the Second World War the Italian environmental protection movement began to establish itself around several national organizations whose actions were chiefly influenced by the conservationist approach. The first to be established was Italia Nostra on October 29, 1955. Giorgio Bassani wrote that the organization's goal was to "promote democracy in Italy, also through safeguarding its artistic and natural heritage."[8] Italia Nostra's first years were distinguished by battles against the gutting of historic city centers and destruction of artistic heritage, as well as by subsequent campaigns in defense of urban greenspaces and natural landscapes.

Between 1959 and 1966 three more associations were created whose principal objective was to promote the safeguarding of natural environments through the creation of "protected areas" (e.g., national parks, reserves, protected wild areas): the federation Pro Natura Italica (1959), which brought together several local associations interested predominantly in nature;[9] the Lega Nazionale contro la Distruzione degli Uccelli (1965, subsequently Lega Italiana Protezione Uccelli [LIPU]), and the Italian division of the World Wildlife Fund (WWF, 1966).[10]

These associations— actually expressions of narrow sectors of the progressive bourgeoisie—found themselves operating in a social and political context fundamentally indifferent to the environmental problems caused by urbanization and industrialization. In the years of rapid transition from prewar scarcity to the postwar "economic miracle," environmental questions were granted very little space within a political and cultural debate that appeared to be dominated by the economy, considered essential to the pursuit of an increasingly widespread "well-being." As Mario Diani observed, there was a total absence of a "cultural orientation favorable to environmental defense that had any mass consistency behind it," and furthermore, in those associations with a national organizational structure, there was an almost total lack of "basic action on the part of citizens."[11]

It was only in the first half of the 1970s, in the wake of the nascent debate on the "stationary society" and the finiteness of natural resources[12] that the term *ecology* began to appear in public debate.[13] In 1971 the proceedings of the Senate Steering Committee for Ecological Problems were published, in which the president of the National Research Council, Vincenzo Caglioti, spoke of "limited resources" to be administered with "prudence and rationality," concluding with: the "future of humanity" is "closely linked to the most rational utilization of natural resources, to elimination of the various forms of pollution and to the availability of suitable means for regulating demographic development."[14] Between 1972 and 1973 certain newspapers (*Il Corriere della Sera*, *La Stampa*, *Il Mondo*, and *L'Espresso*) published a series of articles dedicated to a report promoted by the Club di Roma and carried out by the System Dynamics Group of MIT headed by Donella and Dennis Meadows: *The Limits to Growth: A Report for the Club of Rome's Project on the Predicament of Mankind* (*I limiti dello sviluppo*).[15] Published in 1973 the *Prima relazione sulla situazione ambientale del Paese* still remains one of the most important documents on the deterioration of Italian environmental systems.[16] Then in 1976 an accident at the chemical factory ICMESA at Seveso released a cloud of dioxin which covered over an 18-km^2 area. The accident caught the attention of the media, which led to a broad environmentalist mobilization—considered by many to be a fundamental milestone in the history of, not solely Italian, environmentalism.[17]

In parallel with the publication of environmental crises, the conceptual foundations of environmentalism were radically transformed by the integration of scientific ecology into political and social theory. Political ecology—whose birth is generally agreed to be the publication of Rachel Carson's *Silent Spring* and Barry Commoner's *The Closing Circle*[18]—considered environmental problems as closely correlated to industrial and urban development and saw in the nascent environmental movement a social force capable of influencing the political debate through pressing demands to go beyond the model of capitalistic development.

In Italy the rationale behind political ecology—reprised chiefly from certain heterodox Marxist intellectuals[19]—slowly coalesced around the first battles for health in the workplace, with the action of antagonistic social movements, and with the birth of a few professional organizations aware of environmental problems, such as Medicina Democratica and Geologia Democratica.[20] Against this background environmental associationism began to advance and linked with logistics of a political nature.

The Amici della Terra[21] was founded in 1977 and, together with the Antihunting League (LAC) and the Antivivisection League (LAV), represented an environmental pressure group associated with the Radical Party. The Amici della Terra was immediately and particularly active in the promotion of alternative energy policies and was against the National Energy Plan's decision in favor of nuclear energy: twenty new power plants were to be built (over and above the one already under construction in Caorso, Piacenza Province).

The League for the Environment (subsequently named Legambiente) was set up in 1980 within the Italian Recreational and Cultural Association (ARCI) and right from the start showed itself to be a "flexible" organization dealing with multiple environmental issues, ranging from town planning to the conservation of natural resources, from antipollution to the promotion of alternative lifestyles.[22]

The following year, during a meeting organized in Bologna by the magazine *La nuova ecologia*, saw the creation of the ecological coordination body Arcipelago Verde,[23] a network of associations and local groups whose shared cultural base adhered to the principles of political ecology. The coordination body met periodically in Bologna to deal with themes such as the antinuclear struggle, consumer defense, and ecological lifestyles.

The advent of political ecology organizations spurred the conservationist associations to broaden their actions to include such themes as criticism of the industrialist development model and the crisis of natural and energy resources. For example, the WWF began to fight against pollution and the use of nuclear energy, while Italia Nostra engaged in campaigns for the protection of surface waters and the replacement of naphtha with diesel fuel for heating buildings in urban areas.[24]

The organizational transformation of these associations and the progressive worsening of ecological problems affecting urban areas (e.g., water and air pollution, lack of public greenspaces, and the building of suburbs without the most elementary sanitary infrastructure) convinced an increasing number of people to back the environmental movement: at the beginning of the 1980s the four main Italian associations (WWF, Lega per l'ambiente, LIPU, and Italia Nostra) had a combined membership of 76,000.[25]

In the course of the second half of the decade the environmental movement expanded further thanks to the nuclear question, which between the late 1970s and early 1980s had already favored a collaboration among conservationist, environmental, political, and social groups. This turned into a mass mobilization in favor of the first referendum campaign promoted by Amici della Terra and the Radical Party for abrogation of the regulations for the location of nuclear power plants.[26]

The April 1986 disaster at Chernobyl relaunched the antinuclear movement, backed by the Federazione dei Verdi and the main environmental associations, which called for a referendum by collecting more than one million signatures.[27] The referendums of November 8 and 9, 1987, ratified the abandoning of the nuclear energy plan and marked a clear success for Italian environmentalism, which is reflected in the organizations' growth: around the mid-1990s the WWF reached 281,000 members, Legambiente 115,000, and the Italian branch of Friends of the Earth 25,680.[28] These numbers suggest that the early 1990s was an actual golden age of ecologal associationism.

In these years of greatest political and social popularity, the Italian environmental movement appeared to be hegemonized by organizations of a national character distinguished by a strong central structure and extensive regional ramifications. Predominant among these (at least numerically) were the ones with a conservationist approach, first and foremost the WWF, whose structure, in Diani's view, was the only one that might be compared to the environmental lobbies in English-speaking countries.[29] It was followed by the political ecology organizations, the most significant undoubtedly being the Legambiente which over the years had promoted an important series of campaigns (for lead-free gasoline, for qualitative control of the seas, and the first initiatives for protecting urban ecosystems) to obtain legislative interventions for environmental protection.[30] Then there were militantly active organizations such as Greenpeace Italia (1986) and Mountain Wilderness (1987) whose approach to environmental problems was of a universalistic kind and favored demonstrations to lobbying activities.[31]

This "green archipelago," was completed by the organizations that worked in sectors contiguous to those of the environmental movement (Fondo per l'ambiente Italiano [FAI], Touring Club, Club Alpino Italiano) and by right-wing environmental groups.[32] Simplifying the complex heterogeneity of the ecology galaxy, one might therefore underscore the decided prevalence of a vertical type of organization in which the central structure and its peripheral ramifications run by regular militants were flanked by the wide-ranging baseline participation of volunteers, who helped make the organizational structures less formal.

This model was rejected by local civic committees, single-issue committees, and the political ecology groups—still decidedly minorities within the environmental movement—which were openly critical of bureaucratized associationism and, as Diani has observed, had no intention of "carrying forward into the ecologist commitment the vertical type organizational forms experimented with by the little parties of the new left during the sixties."[33]

THE METAMORPHOSIS: CRISIS OF ENVIRONMENTALISM AND SPREAD OF LOCALIST MOVEMENTS

From the mid-1990s the transformation of the environmental movement was characterized by two phenomena: the progressive institutionalization of environmental organizations and the appearance of numerous protest movements at the local level, which gradually assumed an increasingly particularistic and localistic character. A vast range of literature maintains that expansion of the environmental movement coincided with the transformation of associations into "formal, professional and tangentially centralized" organizations run from great bureaucracies with high levels of professionalization.[34]

This ended up transforming the nature of environmental associationism: the relationship between members and management groups became more labile; an increasingly larger part of the annual budget was spent on marketing, with the aim of convincing people to renew their membership from one year to the next; and the mass character of these organizations and their relationships with institutions accentuated a willingness to compromise and pushed environmental protest toward more and more moderate forms.

The moderation of the main environmental organizations became an indispensable condition for continuing to receive financing from institutions and for retaining that substantial number of members with little inclination to back not only forms of protest considered radical (e.g., nonviolent action such as hunger strikes) but also those of a demonstrative nature (e.g., marches and other public protests in the streets).[35]

The incremental institutionalization of environmental associations—their having become nongovernmental organizations with a structure very similar to that of a traditional political party—led a growing number of citizens to distance themselves from those organizations which comprised the formal environmental movement[36] and seek alternative forms of protest.

These alternative forms, in general organized by civic committees, have been interpreted as basic participatory experiences assimilable to the environmental justice movement[37] or as conservationist stances—generally indicated by the acronyms LULU (Locally Unwanted Land Uses, neutral from an evaluative viewpoint) and NIMBY (Not in My Back Yard, whose accepted meaning has negative connotations)—dictated by particularistic and egoistic motivations.[38] The acronyms LULU and NIMBY are usually employed to indicate popular protests of an environmental nature against the building of infrastructure of public interest (highways, high-speed railways, garbage dumps, waste-to-energy plants, liquid natural gas [LNG] terminals). Often these movements are referred to by speaking generically of the *NIMBY Syndrome*, in which their dissent is not due to unawareness of the social utility of the infrastructure in question, but only that they want it located elsewhere (hence Not in My Back Yard).

During the first decade of the twenty-first century, the phenomenon of local opposition to the building of infrastructure involved numerous European states, taking on notable proportions above all in Italy, where the main reasons behind the explosion of NIMBYism are traceable not only to the fear of pernicious consequences to health and the environment but the worry about unfair distribution of costs and benefits (the former being chiefly borne by local communities), inadequate involvement of citizens in the decision-making process (in France for example there is an "independent commissioner," while in the United states the instrument of "public inquiry" is employed), and mistrust of the political parties and the system of representation.[39] For example, participation of the Green Party in alliances of the center-left and the left ended up weakening the innovative charge of an environmental-political proposal by driving a considerable part of their electorate toward abstention or new, more grass-rooted proposals of both the association (e.g., the Rete dei Comitati per la Difesa del Territorio,[40] promoted by Alberto Asor Rosa) and the political types (e.g., the Cinque Stelle [Five Stars] civic lists promoted by Beppe Grillo).

According to Alessandro Beulcke, NIMBYism has become a typically Italian phenomenon: in the climate of permanent electoral campaign that characterized Italy in the early 2000s, the political parties appropriated NIMBY protests in order to obtain approval. Conflicts were no longer limited to opposition between citizens and the promoters of the works but were extended to local majorities and oppositions, or local administrations and central government, ending up by stoking long, contentious proceedings.[41] Between 2004–2005 (the time span of the first monitoring organized by the NIMBY Forum) and 2011, civic committees became considerably widespread. The second Osservatorio NIMBY Forum (2005–2006) reports 173 episodes of opposition, distributed chiefly in the north (58.96% compared to 21.39% in the south and 19.65% in central Italy). These protests were against: waste management plants (53.76% of the total), production and distribution of energy from nonrenewable sources (26.01%), construction of transportation infrastructure (12.14%), production of electricity from renewable sources (7.51%), and construction of other infrastructures (0.58%).[42] The seventh meeting of the Osservatorio NIMBY Forum (2010–2011) highlighted 344 episodes of opposition, with an increase of 98.84% over the second meeting. These protests involved plants still located mainly in the northern regions (47.38%), but it should be underscored that there was a notable increase of episodes concerning plants in the central regions (26.74%, with 25.88% in the south).[43]

The most important novelty was that the civic committees mainly directed their action toward structures considered indispensable for sustainable development and for respecting the objectives imposed by international treaties, such as the Protocol for the Reduction of Greenhouse Gas Emissions (Kyoto Protocol). In 2010–2011 the protests concerned power plants using renewable sources (45.64% of the total), plants for waste management (27.33%), production and distribution of energy from nonrenewable sources (20.64%), construction of transportation infrastructure (4.07%), and construction of other infrastructure (2.32%).[44] The datum on power plants using renewable sources

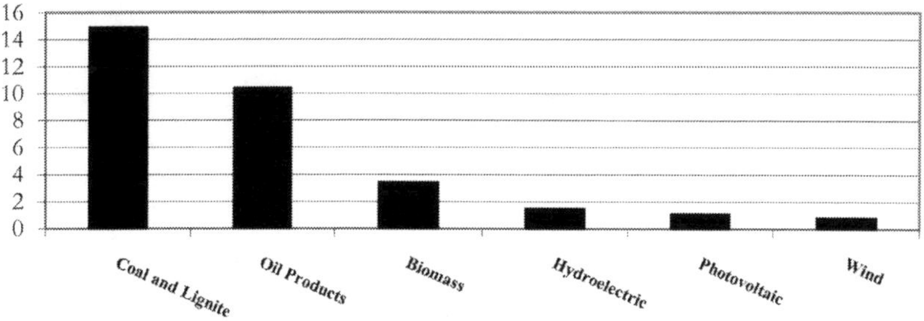

Graph 9.1. Estimates of external costs (Eurocents/kWh) associated with the production of electricity. *Source*: Pietro Menna, *L'energia pulita: Sole, acqua, vento...le fonti di energia rinnovabili* (Bologna: il Mulino, 2003), 85.

is particularly striking (Graph 9.1), as these systems are held to be the energy sources with the lowest externalities (meaning that the environmental costs of the plants are not paid by the producer, such as greenhouse gas emissions from thermoelectric power plants or the risk of radioactive material leakage in the case of nuclear power plants).[45]

We are dealing substantially with an ecological paradox that contributes to introducing the features of civic committees. The first is represented by the attitude to environmental organizations: generally the committees show themselves to be prejudiced against the realization of infrastructure proposed by political policy makers and appear willing to dialogue only with those organizations of the formalized movement which closely align with their own positions. This produced an incremental distancing of the committees from the main political ecology organization, Legambiente, which is the most coherent supporter of the green economy (i.e., founded on renewable energy) and is favorable—when all safety conditions are met—to the construction of plants indispensable to the management of important ecological problems, such as the disposal of solid urban waste. The stance of the committees has moreover contributed to dividing (and weakening) the environmental movement by revealing the divergences among the various associations, particularly between Legambiente and the ones more linked to conservationist ideologies, such as Italia Nostra and the WWF: the former usually comes out against the creation of wind farms while the latter is particularly critical of LNG terminals and waste treatment plants.[46]

The second feature is represented by the strong ideologies of civic committees (or of their internal leadership). These localistic movements are grafted onto a tradition—that of the small political ecology groups which developed in the 1970s around radical left party circles—in which the ideological element has always played a central role. The prevalent ideological references seem to be anti-utilitarianism,[47] organicism,[48] deep ecology,[49] and bioregionalism.[50] These cultural attitudes—being by nature little inclined to retreat from a strictly ideological analysis of the problems imposed by the social, economic, and political dynamics of postcapitalist societies—often employ concepts as inaccurate as they are fundamentalist, such as the idea of "de-growth," referred to by many movements that are against renewable energy plants. They are probably ignorant of the fact that Latouche himself indicates the solar economy as essential to the passage "from an economy of depredation to an ecosustainable system."[51] Radical ideologies, then, seem to be used to build ideological barriers, with the explicit desire to impede (or limit) any constructive dialogue with the formalized environmental organizations and with policy makers. This reveals the egoistic nature, typically NIMBY, that characterizes a significant number of civic committees which, manipulatively waving new ideological totems, end up slowing down those political choices indicated at the local level—as well as at national and transnational levels—as being the only ones possible for setting in motion the (albeit unsure) solutions to problems such as the constant increase in greenhouse gas emissions or the presence in urban areas of dangerous pollutants.

* * *

The reasons behind the progressive success of civic committees that oppose public interest infrastructure seem multiple and complex. One of these is the fact that generally the population has not been involved in the decision-making process, which is handled at the institutional level first by the central government and then by the regional administrations, together with local bodies. This citizen noninvolvement has ended up negatively influencing the perception of risk connected with a new plant; in fact, very often protests have been fed by scarce knowledge due to a lack of transparency and the absence of suitable public information campaigns by institutions.[52]

A further cause lies in the unequal distribution of costs and benefits: very often the people are opposed to an infrastructure or a plant precisely because they have judged the possible benefits as not being worth the negative consequences (environmental impact, health risks, worsening quality of life). A final reason for the proliferation of civic committees lies in the fact that they have been perceived by the population as an efficient expression of direct democracy, whereas environmental organizations, on a par with the Greens, have been (and are) considered too bureaucratic, excessively liable to compromise, and not very rooted in the area. In a word, the committees seem to be the product of the fracture created between environmental organizations and a new model of environmentalism grounded in bottom-up participation and themes of social justice.

The committees—constituted as they are of many different and sometimes also conflicting souls—appear to be highly multifaceted and complex entities which are not possible, without risking oversimplification, to put on the same plane by labeling them indistinctly as NIMBY. A large part is indubitably driven by NIMBY reasons, while others are also driven by motivations of a more general nature, such as the proposal of alternative models for the management of environmental policies (e.g., cold disposal of waste instead of incineration). Still others aspire to the proposal of an alternative development model.[53]

There remains one final (but not secondary) question. Do these forms of participation—whose matrices in Italy still remain very far from structured participatory apparatuses, such as the *orçamento participativo* (participatory council) in Brazil or the Commission Nationale du Débat Public in France[54]—represent new tools capable of strengthening the democratic game, or are they instead demagogic and neo-populist forms of political action?

The civic committees are part of movements which tend to set the "real will of the people" against that of legitimately elected institutions, and in so doing end up by delegitimizing representative democracy—seen as an expression of oligarchies that act against the interest of citizens—against which they oppose the virtues of direct democracy. To define such movements Michelangelo Bovero has used the term *antidemocracy* instead of *antipolitics*,[55] since their distancing from organizational and political representation ends up casting doubt on the very legitimacy of institutions by distorting their role and devaluing their importance in maintaining a correct democratic dialectic. This question should not be underestimated because, as Tony Judt admonished, the state is "the only forum in which it is possible to practice politics," understood as the "appropriate form in which to express different collective preferences in open societies."[56] Civic committees refute de facto the role of the state and local institutions by not recognizing the validity of decisions approved by representative assemblies, claiming the "real will of the people." I am talking about an actual overturning of the principles of representative democracy whose necessity was considered by Judt as "the best argument in favor of the state" since "the two things are destined to live or die together."[57]

In conclusion, the hegemonization of environmentalism by localist movements appears to produce two consequences. The first is the progressive deterioration of the democratic game, since the will of the people expressed through electoral mechanisms, exercised by means of representative assemblies and safeguarded by the constitution, is de facto subordinated to the will of small minorities such as civic committees.[58] The second is the transformation of environmentalism which, as expressed by localist movements, ends up losing its innovative charge and global attitude and being restricted to the role of a social player for the defense of egoistic interests or uncritical support of highly ideologized and fundamentalist theories.

CHAPTER 10

POPULAR INITIATIVE MOVEMENTS AND THE COMMONS

PROMOTION OF AN ALTERNATIVE ECONOMIC MODEL OR DEFENSE OF THE PARTICULAR?

The present moment seems characterized by a crisis of democratic systems consequent to certain transformations of contemporary societies: the tearing of the social fabric caused by increasingly marked inequalities and the multiplication of conflicts;[1] the loss of traditional identities with the waning of the great nineteenth- and twentieth-century narratives and the Fordist work system; the progressive loss of states' sovereignty as an effect of globalization processes and the power acquired by transnational economic players in the context of a neoliberal capitalism which, as Francesco Tuccari efficaciously summarized, "has brought about actual earthquakes" through a "total commodification" of labor "by activating . . . on a local and global scale, merciless competition between half a billion workers, who for decades enjoyed rights and good wages, and one and a half billion new earners who work under inconceivable conditions and for miserly wages."[2]

The product of these processes was (and is) the progressive deterioration of the quality of free democracy which has begun to change toward new forms of leadership by "politicians equipped with a strong brand identity" committed to recounting (in the name of the people and for the people) seductive stories "capable of nourishing the ravenous agenda of the media,"[3] or toward "headless democracies" dominated by "impersonal, anonymous, invisible and politically irresponsible powers . . . infinitely more dangerous since they answer to no one."[4]

THE CRISIS OF REPRESENTATIVE DEMOCRACY AND THE GROWTH OF LULU MOVEMENTS

The crisis of representation has concerned not only political players but also intermediate bodies such as, to keep to our subject, environmental organizations. Since the mid-1990s transformation of the environmental movement has been characterized by two phenomena: the progressive institutionalization of environmental organizations and the appearance of numerous protest movements at the local level which have gradually assumed an increasingly particularistic and localistic nature. A vast range of literature maintains that expansion of the environmental movement coincided with the transformation of associations into "formal, professional and tangentially centralized" organizations run by great bureaucracies with high levels of professionalization.[5]

This transformed the nature of environmental associationism: the relationship between members and management groups became more labile; an increasingly larger part of the annual budgets went to marketing, with the aim of convincing people to renew their membership from one year to the next; the mass character of these organizations and their relationships with institutions wound up accentuating a willingness to compromise and pushed environmental protest toward increasingly moderate forms. The moderation of the main environmental organizations became an indispensable condition for continuing to receive institutional financing and for retaining that substantial number of members who have little inclination to back not only forms of protest considered radical (nonviolent actions such as hunger strikes) but also those of a demonstrative nature (e.g., marches and other protest demonstrations).[6] The progressive institutionalization of environmental associations, their having become nongovernmental organizations with a structure very similar to that of a traditional political party, led a growing number of citizens to distance themselves from these organizations and to seek alternative forms of protest. In this context—characterized by crises in the rules of the democratic game and by the reflux of associative environmentalism—a growing request for participation emerged which was concretized in the birth of numerous movements of citizens ostensibly committed to the defense of "the commons" against the dominant economic model.[7]

DEFENSE OF THE COMMONS OR OF PARTICULAR INTERESTS?

With the assumption that the crisis of representative democracy and the importance of the problems related to management of the land and shared resources have determined the need for new forms of basic participation, the question is whether these movements represent tools capable of reinforcing the democratic game or instead constitute a part of the problem itself, which is to say, contributors to the interruption of democratic political decision making.

Analyzing certain texts often cited within LULU committee networks, we learn that commons are considered "special resources" that possess two "essential" features. "The first: no one can claim to have produced them by himself: the atmosphere and the climate, free or fossil solar energy, the waters and mineral deposits, animals and wild plants, germplasm and seeds and any other form of life capable of reproducing itself spontaneously are gifts of creation, treasures we have found in the magic bucket from which the rainbow originates, exactly like knowledge, culture, languages, codes, scientific acquisitions, artistic heritage and every other fruit of social creativity. The second feature: they are necessary, indispensable and irreplaceable resources for the life of each individual."[8]

In this picture the commons become "the fulcrum around which to retie environmental and social policies" to be managed by means of "an interpersonal relationship and a rapport of cooperation, solidarity and sharing which is denied by the logic of the market of goods."[9] The supporters of this approach maintain that the commons are not monetizable and for this reason they criticize the economic-environmental approach[10] inasmuch as "flanking a monetary value with a natural or cognitive or affective and sentimental resource . . . is highly dangerous since it may suggest that it is possible to buy and sell everything." Nonmonetizable goods include "shared resources and services that are natural, tangible, exhaustible; immaterial, cognitive, limitless shared resources and services; public natural and artificial resources and services" (e.g., physical and digital infrastructures, welfare, the Internet) as well as "global shared resources (atmosphere, oceans, forests, biodiversity)" and "shared resources linked to territorialized civic use (local commons such as hydrographic basins, urban ecosystems)."[11] To defend the commons against commodification a "decisive revolutionary war aimed at survival of the planet" is hoped for in which the "dominant paradigm founded on a Darwinist idea of the world" shall be countered by the "recessive one founded instead on an ecological and community idea of the world." The statement of the dominant model, characterized by the "rhetoric" of growth and development, abandons the community model as a consequence of the "gradual outcome of markets' needs to be founded on political institutions of State dimension." The opposition to this model is identified in the "periphery" where the recessive model "seeks to resist strenuously the merciless assault consisting of structural (World Bank plans . . ., privatization of water) and cultural adjustment."[12] The recovery of the commons may therefore come about

> within the framework of a deep clash—epistemological and also psychological—between two worldviews (the mechanistic-technological, founded on individualism, domination and the quantitative dimension, and the ecological, founded on community, holism and the qualitative dimension). A clash that should be translated into a revolutionary political praxis capable of bringing about the global triumph of the latter in an extremely brief space of time, since it is the only scientific conception compatible with the long-term maintenance and adaptation of life on our planet. We are dealing substantially with setting up a political and cultural alternative capable of unseating both private property and state sovereignty from the role of cornerstones of the political organization that are exhaustive of the range of possibilities.[13]

In this context the committees against locally unwanted land uses may consider themselves one of the tools for concretizing the "revolutionary war" against the dominant model. However, if the objective of the "recessive paradigm" is that of defending a "natural exhaustible commons" and the "global commons," the practice does not appear to follow the theory. In fact we have seen that the majority of opposition is against infrastructure which science holds to be indispensable precisely for defending the global and local commons against the effects of climate change (considered the chief planetary environmental problem, at the center of the recent Conference of the Parties held in Paris from November 30 to December 11, 2015). In a particular way, the darts of the *benicomunisti* (commons-ists) are aimed against the production of electricity from renewable sources, whose development is considered indispensable for reducing emissions from the so-called coal economy. Not even public transportation systems remain immune to protest, although they are necessary for creating a mobility alternative to that of private transportation on the roads and, consequently, for protecting common resources linked to territorialized civic use (e.g., air quality within urban ecosystems). In this case we are faced with an evident contradiction between the committees' theoretical premises and their actions.

The extent of the antilogy between theory and practice becomes clear if we analyze the protests, using two cases I have been following for some years. The first (2005–2008) concerns a wind farm in Poggi Alti in the municipality of Scansano, Grosseto Province (ten wind turbines for an overall yield of 20 MW). Opposition to the plant—mainly contesting the visual impact of the wind towers on the landscape—was organized by citizens' committees led by the winegrowers of the area (where Morellino is produced, one of Tuscany's most famous wines) and Italia Nostra, while Legambiente, the WWF, and Greenpeace expressed a substantially favorable opinion. The national press sided with the citizens' committees, and an influential lead writer for the daily *La Repubblica* defined the affair as "a crazy screenplay of Don Quixote with the monstrous windmills assaulting the castle of Dulcinea"[14] (meaning that of Montepò, owned by the winegrowing family Biondi-Santi, in the vicinity of the wind farm).[15] The second (under way for more than ten years) is directed against the construction of a new rail network in Florence, described as a step toward destruction of the city. The protests concentrated on the cutting down of trees along the course of the tramway and the potential damage that trams would cause to the city's main historical-artistic heritage (the cupola of Brunelleschi, the campanile of Giotto, the Baptistery of San Giovanni). On February 17, 2008, an advisory referendum was held (in which 39.36% of eligible voters took part) which confirmed the existence of substantial opposition to the project, with 52.85% voting against the referendum.[16]

In both cases it is easy to highlight the antilogies (e.g., wind power has a very low impact in comparison to conventional technologies; mobility by rail is the most efficient solution to reducing traffic pollution). Nor is it difficult to pick out the purposes in defending their particular interests: first, economic activities managed by the promoters of the protest,[17] and second, the reaffirmation of motorized individualism (i.e., the freedom to travel with one's own two- or four-wheeled vehicles, threatened by new infrastructure that reduces parking space and otherwise creates practical conditions favorable to trams and disadvantageous to motor vehicles).[18]

The LULU committees, far from being the standard-bearers of the "decisive revolutionary war," end up feeding a series of ecological paradoxes that reveal peculiar features. The first is the attitude of closure assumed by environmental associations: generally the committees show themselves to be prejudicial to the creation of the infrastructure proposed by political decision makers and appear willing to dialogue with only those organizations of the formalized environmental movement who side with positions very close to those of the committees themselves.

The second is their strong ideologies. These localistic movements are grafted onto a tradition—that of the small political ecology groups that developed in the 1970s around radical left party circles—in which ideological dogmatisms have always played a central role. So the ideological element seems to be utilized for raising barriers, with the explicit desire to impede (or limit) any constructive dialogue with the formalized environmental organizations and with policy makers, revealing the egoistic nature that characterizes a significant number of civic committees. Manipulatively waving new ideological totems, they slow down at the local level those political choices most appropriate for solving certain, albeit complex, problems, such as the constant increase in greenhouse gas emissions or the presence in urban areas of pollutants dangerous to health.

These attitudes (assimilable to those of radical environmentalism) have not escaped the more attentive observers. The political analyst James R. Flynn speaks of the "most alarmed environmentalists" as "people ingenuous in politics and in economics,"[19] while anthropologist Bruno Latour extends the criticism to ecological thought:

> The ecologists have not created an artistic and intellectual work to articulate the cause they are struggling for. It's unbelievable: notwithstanding the importance of what is at stake, as great as the social and labor questions of the nineteenth century, there has not been and there still isn't a corresponding theoretical elaboration. Nothing similar to Marxism for example. If we compare it with the theoretical work of two centuries on social injustice, it seems incredible that so little has been done so far for the ecological question. Without complaining about ecology, we need however to acknowledge that it has not been capable of drawing up new concepts but rather has borrowed concepts from other theories and combined them.[20]

There is a further element in Italy that is often underestimated. A large segment of the environmental movement—for example, the committees against photovoltaic plants, wind farms, and rail/tram infrastructure, or movements such as the Rete dei Comitati per la Difesa del Territorio (Network of Territory Defense Committees) promoted by Alberto Asor Rosa in 2006,[21] as well as some associations like Italia Nostra and FAI—when speaking about the environment, are not referring to the natural environment as the "grouping of physical, chemical and biological conditions" that characterize an "open system within which there are exchanges of energy and information and which is capable of self-regulation and the maintaining of dynamic equilibrium."[22] Instead, they refer to the landscape, which has been radically modified by man over the course of time. The reference, in brief, is to a very precise idea of nature[23] which does not at all correspond to the natural environment but to places (e.g., Langhe, Chianti, Maremma, areas containing several archaeological parks) that have been ably molded by man and, precisely for this reason, are considered the expression of an aesthetic (of renaissance, romantic, and neoclassical derivation) that must be preserved. Furthermore, the idea that it is necessary to maintain some of these landscapes unaltered (think Langhe and Chianti) is also inspired by their high economic value, because in the world's eye they identify with the Italian agricultural sector (wine growing, olive growing, DOP[24] food products, and truffles). In a word, we are talking about areas where the "recessive paradigm" is certainly not experienced and which by now have very little left of the natural (in the sense of nonanthropogenic), and also because the activities practiced there produce impacts on local natural resources (e.g., the use of mechanized farming, fertilizers, pesticides) and global ones (tourism has a high carbon footprint and generates a considerable quantity of waste whose disposal constitutes an externality to the detriment of local communities) that are decidedly greater than those of a wind farm or photovoltaic power plant.

The action of these movements therefore appears quite out of line with the "decisive revolutionary war" in favor of the commons, rather involving protectionism of multiple particular interests. These include defense of the supposed supremacy of the classical and modern age[25] over the contemporary, promulgated by conservationism of a traditionalist matrix à la Cederna (somewhat elitist, somewhat fundamentalist). This produces results opposite to what it claims to seek, because its opposition to the most innovative solutions continues to encourage the maintenance of the status quo, thus contributing to the failure of land management policies.

In light of the foregoing considerations I am led to consider the LULU movements (which, it will be recalled, now enjoy a wide range of literature in their favor)[26] as an element that contributes both to weakening the environmental movement (by confining it to the role of social player seeking to defend particular interests) and to interrupting the democratic circuit of policy making, contributing to the subordination of popular will, expressed through electoral mechanisms, to that of small minorities.[27]

The civic committees, by setting the "real will of the people" against that of legitimately elected institutions, end up by delegitimizing representative democracy, to which they oppose the virtues of direct democracy. To define such movements Michelangelo Bovero has used the term *antidemocracy* instead of *antipolitics*,[28] since the distancing from associative and political representation ends up casting doubt on the very legitimacy of institutions (considered the tools of oligarchies that act against the interest of citizens) by distorting their role and devaluing their importance in maintaining a correct democratic dialectic. The question should not be underestimated because, as Tony Judt admonished, the state is "the only forum in which it is possible to practice politics," understood as the "appropriate forum in which to express different collective preferences in open societies."[29] Civic committees, by not recognizing the validity of decisions approved by representative assemblies, refute both the role of the state and of local institutions. We are dealing with an actual overturning of the principles of representative democracy, whose necessity was considered by Judt "the best argument in favor of the state" given that "the two things are destined to live or die together."[30]

In the field of juridical science there are some who express perplexity about "overenthusiastic interpretations of environmental participation," judging them as being exposed to a twofold risk:

> A simplistic approach to participatory themes does not permit in the first place recording and consequently assessing the disparities of position connoted by the subjects called upon to participate. This leads to an underestimation of the way in which these disparities in turn have repercussions on carrying out the participatory processes and on their possible outcomes. . . . This risk of simplification is moreover accompanied by a problem of ineffectiveness of the participatory tools made available by judicial norms. Conversion into formalized devices may turn out to be wholly unproductive at the level of concretely effected contributions, in cases in which it is not supported by previous consideration of the participants' unequal initial conditions. . . . The difficulty does not lie in the introduction of any kind of participative expedient but rather in ensuring that it does not, for modalities of implementation and for the subjects to which it is addressed, remain available solely to certain subjects and those with precedents. This consideration is all the more important if related to the question of the environment as a shared resource, since in this case it is on the nature of the interests introduced into the public decision-making processes that the very configurability of the resource in terms of communality depends. Insertion into the decision-making processes of sectorial and particularistic interests, and above all their ability to determine the final results thereof, denaturize the resource in question, annulling its character of "commonality." We thus have the paradox by which participation, from being an instrument for highlighting community practices in the use of resources, becomes a go-between for the dissimulated declaration of interests common to only a few.[31]

To conclude this chapter (well aware that in these few pages it is impossible to deal exhaustively with such a complex and controversial question) I take the liberty of pointing out how the flattening out of historical-environmental reflection on the ecocentric positions of localistic environmentalism—professed (not rarely with a display of "blind pride" and "empty formulas"; I refer, for example, to theoreticians of de-growth or primitivism, highly popular within the LULU movements) by theoreticians and citizens convinced of "already knowing every law of things"—risks derailing scientific discussion onto the dead-end track of sterile polarization between supporters of the ecocentric approach and its critics.

NOTES

INTRODUCTION

1. In Italy there is still very little urban environmental historiography. See Gabriella Corona, "Risorse nella città: Natura e territorio a Napoli tra Otto e Novecento," in Piero Bevilacqua and Gabriella Corona (eds.), *Ambiente e risorse nel Mezzogiorno contemporaneo* (Rome: Donzelli, 2000), 191–208; Gabriella Corona, "La sostenibilità urbana a Napoli: Caratteri strutturali e dinamiche storiche," *Meridiana: Rivista di storia e scienze sociali* 42 (2001): 15–43; Simone Neri Serneri, "Uso e abuso delle risorse: L'emergere della questione ambientale," in Simone Neri Serneri (ed.), *Storia del territorio e storia dell'ambiente: La Toscana contemporanea* (Milan: FrancoAngeli, 2002), 101–16; Gabriella Corona and Simone Neri Serneri, *Storia e ambiente: Città, risorse e territori nell'Italia contemporanea* (Rome: Carocci, 2007); Gabriella Corona, *I ragazzi del piano: Napoli e le ragioni dell'ambientalismo urbano* (Rome: Donzelli, 2007); Vanni Bulgarelli and Catia Mazzeri, *La città e l'ambiente: Le trasformazioni ambientali urbane a Modena nel Novecento* (Modena: Apm Edizioni, 2009); Anna Dell'Aquila, *Caserta (1945–1974): Una storia urbana e ambientale* (Arezzo: Fruska, 2013); Federico Paolini, *Firenze 1946–2005: Una storia urbana e ambientale* (Milan: FrancoAngeli, 2014); Giacomo Parrinello, *Fault Lines: Earthquakes and Urbanism in Modern Italy* (Oxford: Berghahn Books, 2015).

2. Salvatore Adorno and Simone Neri Serneri, "Per una storia ambientale delle aree industriali in Italia: Introduzione," in Salvatore Adorno and Simone Neri Serneri (eds.), *Industria, ambiente e territorio* (Bologna: il Mulino, 2009), 25.

3. Italian historiography on industrial environmental impacts is still marginal and deals prevalently with accidents and pollution. See Marino Ruzzenenti, *Un secolo di cloro e . . . PCB: Storia delle industrie Caffaro di Brescia* (Milan: Jaca Book, 2001); Simone Neri Serneri, *Incorporare la natura: Storie ambientali del Novecento* (Rome: Carocci, 2005); Laura Centemeri, *Ritorno a Seveso: Il danno ambientale, il suo riconoscimento, la sua riparazione* (Milan: Bruno Mondadori, 2006); Salvatore Adorno and Simone Neri Serneri (eds.), *Industria, ambiente e territorio: Per una storia ambientale delle aree industriali in Italia* (Bologna: il Mulino, 2009); Bruno Ziglioli, *La mina vagante: Il disastro di Seveso e la solidarietà nazionale* (Milan: FrancoAngeli, 2010); Bruno Ziglioli, *Sembrava nevicasse: La Eternit di Casale Monferrato e la Fibronit di Broni; Due comunità di fronte all'amianto* (Milan: FrancoAngeli, 2016).

4. Bruno Latour, *Un monde pluriel mais commun: Entretiens avec François Ewald* (La Tour d'Aigues: éditions de l'Aube, 2003); Clive Hamilton, *Requiem for a Species: Why We Resist the Truth about Climate Change* (London: Earthscan, 2010), 66–158.

5. In Italy no historiography exists to date on the environmental impacts of consumption. International historiography is also exiguous: searching for the word *consumption* in the main historical-environmental database (the Forest History Society, https://foresthistory.org, accessed Feb. 24, 2018) produced 110 hits. Those concerning models of consumption, however, were limited to the following: Alan Thein Durning, *How Much Is Enough? The Consumer Society and the Future of the Earth* (New York: W. W. Norton, 1992); Thomas Princen, Michael Maniates, and Ken Conca (eds.), *Confronting Consumption* (Cambridge, MA: MIT Press, 2002); Worldwatch Institute, *State of the World 2004: Special Focus, The Consumer Society* (New York: Worldwatch Institute, 2004); Peter Dauvergne, *The Shadows of Consumption: Consequences for the Global Environment* (Cambridge, MA: MIT Press, 2008); Monika Dittrich, Stefan Giljum, Stephan Lutter, and Christine Polzin, *Green Economics around the World? Implications of Resource Use for Development and the Environment* (Vienna: Sustainable Europe Research Institute, 2012); Vaclav Smil and Kazuhiko Kobayashi, *Japan's Dietary Transition and Its Impacts* (Cambridge, MA: MIT Press, 2012); Judith Shapiro, *China's Environmental Challenges* (Malden, MA: Polity Press, 2012); Maureen Ogle, *In Meat We Trust: An Unexpected History of Carnivore America* (Boston: Houghton Mifflin Harcourt, 2013); Gay Hawkins, Emily Potter, and Kane Race, *Plastic Water: The Social and Material Life of Bottled Water* (Cambridge, MA: MIT Press, 2015); Alex V. Barnard, *Freegans: Diving into the Wealth of Food Waste in America* (Minneapolis: University of Minnesota Press, 2016); H. Louise Davis, Karyn Pilgrim, and Madhudaya Sinha (eds.), *The Ecopolitics of Consumption: The Food Trade* (Lanham, MD: Lexington Books, 2016).

6. Jared Diamond, *Collapse: How Societies Choose to Fail or Succeed* (London: Penguin Books, 2006).

7. Law No. 319, *Norme per la tutela delle acque dall'inquinamento.* The objectives of the law were to regulate the discharges of any kind in surface, underground, internal, and marine waters. The act elaborated general criteria for water use; regulated the construction of aqueducts, sewers, and water purification plants; drew up a national water rehabilitation plan; and organized a network to control water quality.

8. In the United States urban environmental history has a long tradition. See especially Joel A. Tarr, "Water and Wastes: A Retrospective Assessment of Wastewater Technology in the United States 1800–1932," *Technology and Culture* 25 (Apr. 1984): 226–63; William Cronon, *Nature's Metropolis: Chicago and the Great West* (New York: W.W. Norton, 1991); Joel A. Tarr, *The Search for the Ultimate Sink: Urban Pollution in Historical Perspective* (Akron: University of Ohio Press, 1996); Martin V. Melosi, *The Sanitary City: Urban Infrastructure in America from Colonial Times to the Present* (Baltimore:Johns Hopkins University Press, 2000); Joel A. Tarr (ed.), *Devastation and Renewal: An Environmental History of Pittsburgh and Its Region* (Pittsburgh: University of Pittsburgh Press, 2003); Harold L. Platt, *Shock Cities: The Environmental Trans-*

formation and Reform of Manchester and Chicago (Chicago: University of Chicago Press, 2005); Michael F. Logan, *Desert Cities: The Environmental History of Phoenix and Tucson* (Pittsburgh: University of Pittsburgh Press, 2006); William Deverell and Greg Hise (eds.), *Land of Sunshine: An Environmental History of Metropolitan Los Angeles* (Pittsburgh: University of Pittsburgh Press, 2006); Martin V. Melosi and Joseph A. Pratt (eds.), *Energy Metropolis: An Environmental History of Houston and the Gulf Coast* (Pittsburgh: University of Pittsburgh Press, 2007); Matthew Klingle, *Emerald City: An Environmental History of Seattle* (New Haven, CT: Yale University Press, 2007); David Stradling, *The Nature of New York: An Environmental History of the Empire State* (Ithaca, NY: Cornell University Press, 2010); Char Miller (ed.), *Cities and Nature in the American West* (Reno: University of Nevada Press, 2010); Stephane Castonguay and Matthew D. Evenden (eds.), *Urban Rivers: Remaking Rivers, Cities, and Space in Europe and North America* (Pittsburgh: University of Pittsburgh Press, 2012). As for European historiography, see Christopher Bernhardt (ed.), *Environmental Problems in European Cities in the 19th and 20th Centuries* (Munich: Waxman, 2001); Stephen Mosley, *The Chimney of the World: A History of Smoke Pollution in Victorian and Edwardian Manchester* (Cambridge: White Horse Press, 2001); Christopher Bernhardt and Geneviève Massard-Guilbaud (eds.), *Le démon moderne—The modern Demon: La pollution dans les sociétés urbaines et industrielles d'Europe* (Clemont-Ferrand: Presses Universitaires Blaise Pascal, 2002); André Guillerme, Anne-Cécile Lefort, and Gérard Jigaudon, *Dangereux, insalubries et incommodes: Paysages industriels en Banlieu parisienne XIX-XX Siècles* (Seyssel: Champ Vallon, 2004); Dieter Schott, Bill Luckin, and Geneviève Massard-Guilbaud (eds.), *Resources of the City: Contributions to an Environmental History of Modern Europe* (Aldershot, UK: Ashgate, 2005); Sabine Barles, *L'invention des déchets urbains: France 1790–1970* (Seyssel: Champ Vallon, 2005); Dieter Schott and Michael Toyka-Seid (eds.), *Die europäische Stadt und ihre Umwelt* (Darmstadt: WBG, 2008); Genevieve Massard-Guilbaud and Richard Rodger (eds.), *Environmental and Social Justice in the City: Historical Perspectives* (Cambridge: White Horse Press, 2011).

PART I: NATURAL RESOURCES AND THEIR USES

1. On Tuscany, see Massimo Morisi, Annick Magnier, *Governo del territorio: Il modello Toscana* (Bologna: il Mulino, 2003); Giulio Giovannoni, *Governare il territorio: Una riflessione sul caso toscano* (Milan: FrancoAngeli, 2004); Elena Fasano Guarini, Giuseppe Petralia, and Paolo Pezzino, *Storia della Toscana*, vol. 2, *Dal Settecento ad oggi* (Bari-Roma: Laterza, 2004).

2. Martin V. Melosi, *The Sanitary City: Urban Infrastructure in America from Colonial Times to the Present* (Baltimore: Johns Hopkins University Press, 2000), 283.

3. Gabriella Corona and Simone Neri Serneri, "Città e ambiente nell'Italia contemporanea: Un'introduzione," in Gabriella Corona and Simone Neri Serneri (eds.),

Storia e ambiente: Città, risorse e territori nell'Italia contemporanea (Rome: Carocci, 2007), 11–37. See also Simone Neri Serneri, *Incorporare la natura: Storie ambientali del Novecento* (Rome: Carocci, 2005), 219–33, 234–77.

4. Salvatore Adorno and Simone Neri Serneri, "Per una storia ambientale delle aree industriali in Italia: Introduzione," in Adorno and Neri Serneri, *Industria, ambiente e territorio* (Bologna: il Mulino, 2009), 13–29.

5. The essays in this section are part of an eight-year research project whose final results were published in my book *Firenze 1946–2005: Una storia urbana e ambientale* (Milan: FrancoAngeli, 2014).

6. Corona and Neri Serneri, *Città e ambiente nell'Italia contemporanea*, 18–33.

7. Adorno and Neri Serneri, *Per una storia ambientale delle aree industriali in Italia*, 25–28.

CHAPTER 1: THE ECONOMIC AND ENVIRONMENTAL IMPACTS OF BIOMASS, SOLAR AND MINI-WIND POWER PLANTS

1. World Resources Institute, *World Resources 2005: The Wealth for the Poor; Managing Ecosystems to Fight Poverty* (Washington, DC: World Resources Institute, 2005), 200–202.

2. International Energy Agency (IEA), *World Energy Outlook 2006* (Paris: IEA Publications, 2006), 492.

3. OPEC, *World Oil Outlook 2007* (Vienna: OPEC Secretariat, 2007), 20; Energy Information Administration (EIA), *International Energy Outlook 2008* (Washington, DC: EIA, 2008), 185.

4. IEA, *World Energy Outlook 2008: Sintesi* (Paris: IEA Publications, 2008), 4–5.

5. IEA, *World Energy Outlook 2008: Sintesi*, 5–6.

6. EIA, *International Energy Outlook 2008*, 185, 234.

7. Istituto Nazionale di Statistica (ISTAT), *Statistiche ambientali 2007* (Rome: ISTAT, 2007), 257–77. In 2007 the availability of energy by fuel was: 43% oil, 36% natural gas, 9% coal, 7% renewables, 5% electricity. See ENEA, *Rapporto Energia e Ambiente 2007: Analisi e scenari* (Rome: ENEA, 2008), 25.

8. On the energy situation in Tuscany, see also Agenzia Regionale per la Protezione Ambientale della Toscana (ARPAT), *Relazione sullo stato dell'ambiente in Toscana 2008* (Florence: Edifir, 2008).

9. ENEA, *Libro Bianco per la valorizzazione energetica delle fonti rinnovabili* (Rome: ENEA, 1999), sections 1.3 and 3.2.

10. Unione Petrolifera, *Previsioni di domanda energetica e petrolifera italiana 2007–2020* (Rome, 2007), 2–7.

11. On wind energy, see Richard L. Hill, *Power from Wind: A History of Windmill Technology* (Cambridge: Cambridge University Press, 1994); Alexis Madrigal,

Powering the Dream: The History and Promise of Green Technology (Cambridge, MA: Da Capo Press, 2011).

12. Alberto Mercanti, Rosanna Granatella, and Alessandra La Manna, *Energia eolica: Tra passato e futuro un'alternativa attuale* (Florence: Alinea, 2002), 107–8.

13. Pietro Menna, *L'energia pulita. Sole, acqua, vento . . . le fonti di energia rinnovabili* (Bologna: Il Mulino, 2003), 85.

14. Mercanti, Granatella, and La Manna, *Energia eolica*, 22–26.

15. Legambiente, *Comuni rinnovabili 2008: Rapporto di Legambiente* (Rome, 2008), 24–25.

16. Lester R. Brown, *Piano B 3.0: Mobilitarsi per salvare la civiltà*, trans. Attilio Anitori (Milan: Edizioni Ambiente, 2008), 251–73 (original title, *Plan B 3.0: Mobilizing to Save Civilization*, 2008).

17. Mercanti, Granatella, and La Manna, *Energia eolica*, 89–94; ENEA, *Energia eolica: Aspetti tecnici, ambientali e socio-economici* (Rome: ENEA, 2000); Menna, *L'energia pulita*, 60. See also Mario Zambrini, "La valutazione degli impatti ambientali degli impianti eolici," in Duccio Bianchi and Giuseppe Gamba (eds.), *Ambiente Italia 2007: Rapporto annuale di Legambiente* (Milan: Edizioni Ambiente, 2007), 165–77.

18. Mercanti, Granatella, and La Manna, *Energia eolica*, 105–6.

19. Tommaso La Mantia, Giuseppe Barbera, Rocco Lo Duca, Bruno Massa, and Salvatore Pasta, "Gli impatti degli impianti eolici sulla componente biotica e le misure di mitigazione," in Gianni Silvestrini and Mario Gamberale (eds.), *Eolico: Paesaggio e ambiente* (Rome: Franco Muzzio Editore, 2004), 95–140. The article presents an extensive bibliography on the environmental impact of wind farms.

20. Centro Ornitologico Toscano, *Indagine bibliografica sull'impatto dei parchi eolici sull'avifauna: Relazione e bibliografia commentata* (Florence: Regione Toscana-Giunta Regionale, 2002), 8–9.

21. Regione Toscana, *Piano di indirizzo energetico regionale (PIER): Proposta della Giunta Regionale* (Florence: Regione Toscana, 2008), 62.

22. Regione Toscana, *Piano di indirizzo*, 68.

23. Some case studies are mentioned on the website of the British Wind Energy Association, http://www.renewableuk.com/en/renewable-energy/wind-energy/small-and-mediumscalewind/index.cfm (accessed Feb. 27, 2015).

24. "Trento prova il primo parco minieolico," *La Repubblica*, Sept. 10, 2007.

25. Legambiente, *Proposte per lo sviluppo del minieolico in Italia* (Rome, 2008).

26. Brown, *Piano B 3.0*, 265–68; William Ramsay, *Bioenergy and Economic Development: Planning for Biomass Energy Programs in the Third World* (Boulder, CO: Westview Press, 1985).

27. Ragnar Löfstedt, "Sweden's Biomass Controversy: A Case Study of Communicating Policy Issues," *Environment* 40, no. 4 (1998): 42–45.

28. Menna, *L'energia pulita*, 41.

29. Commission of the European Communities, *Comunication from the Commission: Biomass Action Plan*, Dec. 7, 2005, 20. For 2003 the figure includes firewood in forestry and agricultural waste. Data reported for 2010 and 2030 come from the most prudent hypothesis.

30. See, for example, Worldwatch Institute, *Biofuels for Transport: Global Potential and Implications for Sustainable Agriculture and Energy in the 21th Century* (London: Earthscan, 2007); Andrea Cappelli and Silvano Simoni, "L'oro verde: Vizi e virtù dei biocarburanti," in "Il clima dell'energia; Il tempo che farà, le guerre dei tubi, l'Italia a rischio," *Limes: Rivista italiana di geopolitica* 6 (2007): 145–57.

31. See Greenpeace International, *How the Palm Oil Industry Is Cooking the Climate* (Amsterdam: Greenpeace, 2007); Gregore Pio Lopez and Tara Laan, *Biofuels at What Cost? Government Support for Biodiesel in Malaysia* (Geneva: Global Subsidies Initiative of the International Institute for Sustainable Development, 2008).

32. Mario Conti (ed.), *L'impatto ambientale di centrali elettriche alimentate a biomasse legnose: Esperienze e valutazioni* (Rome: ENEA, 2003), 32. See also the bibliography quoted in this study.

33. Legambiente, *Comuni rinnovabili 2008*, 31–37.

34. Ministero dell'ambiente e della tutela del territorio e del mare, www.minambiente.it (accessed Dec. 20, 2008; no longer available).

35. Conti, *L'impatto ambientale*, 34–35.

36. Regione Toscana, *Rapporto sulle politiche agricole e rurali, allegato 1, cap. 3.1: Analisi della situazione in termini di punti di forza e di debolezza*, supplement to *Bollettino ufficiale della Regione Toscana*, no. 48 (Nov. 28, 2007): 374–87.

37. On solar technology, see Hermann Scheer, *Il solare e l'economia globale: Energia rinnovabile per un futuro sostenibile* (Milan: Edizioni Ambiente, 2004); Frank Laird, *Technology Policy and Institutional Values* (Cambridge: Cambridge University Press, 2001); John Perlin, *From Space to Earth: The Story of Solar Electricity* (Ann Arbor, MI: Aatec Publications, 1999); Daniel M. Berman and John T. O'Connor, *Who Owns the Sun? People, Politics and the Struggle for a Solar Economy* (White River Junction, VT: Chelsea Green Publishing, 1996).

38. Brown, *Piano B 3.0*, 259–64.

39. Brown, *Piano B 3.0*, 259–64.

40. Brown, *Piano B 3.0*, 259–64.

41. Brown, *Piano B 3.0*, 262.

42. The data on photovoltaics are drawn from EPIA and Greenpeace, *Solar Generation V-2008: Solar Electricity for over One Billion People and Two Million Jobs by 2020* (Brussels and Amsterdam: EPIA and Greenpeace, 2008). See also EPIA and Greenpeace, *Solar Generation IV-2007: Solar Electricity for over One Billion People and Two Million Jobs by 2020* (Brussels and Amsterdam: EPIA and Greenpeace, 2007).

43. Brown, *Piano B 3.0*, 261.

44. Xuemei Bai, "Rizhao: Una città a energia solare," in Worldwatch Institute, *State of the World 2007: Il nostro futuro urbanizzato* (Milan: Edizioni Ambiente, 2007), 242–44.

45. For these examples, see Janet L. Sawin and Kristen Hughes, "Dare energia alle città," in Worldwatch Institute, *State of the World 2007*, 228–35.

46. International Energy Agency, Photovoltaic Power Systems Programme, *Compared Assessment of Selected Environmental Indicators of Photovoltaic Electricity in OECD Cities*, IEA PVPS TASK 10, ACTIVITY 4.4., REPORT IEA-PVPS T 10-01:2006, May 2006.

47. On photovoltaic energy in Italy, see Romeo Bassoli, Calogero Messana, and Roberto Vigotti, *Energia dal sole: Prospettive dell'energia fotovoltaica in Italia* (Rome: Hypothesis, 1992).

48. ENEA, *Rapporto energia e ambiente 2005: I dati* (Rome: ENEA, 2005), 250.

49. Bianchi and Gamba, *Ambiente Italia 2007*, 225.

50. The eighteen municipalities were: Prato allo Stelvio (353.9 kW/1,000 residents), Rio di Pusteria (271.4), Fiavè (255.3), San Bassano (242.7), Guardea (195), Crandola Valsassina (193.1), Assoro (187.5), Mestrino (177.7), Laces (172.1), Sant'Agostino (160.7), Ailoche (157.7), Romallo (146.8), Castelletto d'Erro (129.4), Bagnolo Piemonte (128.3), Campo Tures (126.7), Portico (115.9), San Benedetto (115.9), Roccastrada (108.7), and Capriana (103.4). See Legambiente, *Comuni rinnovabili 2008*, 12–15, 16–19.

51. Regione Toscana, *Piano di indirizzo*, 61–62, 65–66, 68; Legambiente, *Comuni rinnovabili 2008*, 5.

CHAPTER 2: THE ENVIRONMENTAL IMPACT OF URBANIZATION AND INDUSTRIALIZATION IN THE GREATER FLORENCE AREA, 1945-2001

This article summarizes research that lasted eight years, the results of which are published in Federico Paolini, *Firenze 1946–2005: Una storia urbana e ambientale* (Milan: FrancoAngeli, 2014).

1. The Florentine area includes the municipalities of Bagno a Ripoli, Calenzano, Campi Bisenzio, Fiesole, Florence, Impruneta, Lastra a Signa, Scandicci, Sesto Fiorentino, and Signa. The municipalities of Cantagallo, Carmignano, Montemurlo, Poggio a Caiano, Prato, Vaiano, and Vernio are part of the Prato area.

2. For a comprehensive account of the developments in the planning of Florence, see Giuseppe Campos Venuti and Odoardo Reali, "Firenze: L'urbanistica contrattata," in Giuseppe Campos Venuti and Federico Oliva (eds.), *Cinquant'anni di urbanistica in Italia 1942–1992* (Roma-Bari: Laterza, 1993), 313–27.

3. In 1951 Florence, Fiesole, Bagno a Ripoli, Impruneta, Scandicci, Lastra a Signa, Signa, Campi Bisenzio, Sesto Fiorentino, Pontassieve, Prato, and Vaglia e Vaiano were part of the PIF. In 1956 Montemurlo and Cantagallo e Vernio joined the PIF; in 1971

Poggio a Caiano, Carmignano, Barberino di Mugello, Greve, and San Casciano Val di Pesa also joined.

4. With the exception of the prefectorial commissioners (1957–1961, 1965–1966, 1969–1970) and of the mayorship of the socialist Lelio Lagorio (1965), the first citizens of the main town of Tuscany had always been appointed by the Christian Democrat Party.

5. Autorità di Bacino del Fiume Arno, *Trasformazioni del territorio e sviluppo dell'edificato lungo il corso dell'Arno e degli affluenti (1954–1993 e 1995)* (Florence: Autorità di Bacino del Fiume Arno, 1997).

6. "Scempio a Monte Morello," *La Nazione*, July 2, 1971.

7. Data refer to the censuses of 1892, 1911, 1927, 1951, 1961, and 1971, reported in Piero Innocenti, *L'industria nell'area fiorentina* (Florence: Associazione degli industriali di Firenze, 1979); and Gianni Lorenzoni, "Lo sviluppo industriale di Prato," in *Storia di Prato* (Prato: Cassa di Risparmi e Depositi, 1980), 3:83–574.

8. These cities increased the proportion of local units in the Florentine area during 1951–1971: Calenzano +460.56%, Scandicci +211.38%, Campi Bisenzio +176.45%, Sesto Fiorentino +121.07%, Signa +94.47%, Impruneta +67.89%, Lastra a Signa +55.23%, Florence +38.02%, Bagno a Ripoli +36.93%, Fiesole +25.47%.

9. On the Florence-Prato metropolitan area, see Lando Bortolotti and Giuseppe De Luca, *Come nasce un'area metropolitana: Firenze, Prato, Pistoia: 1848–2000* (Florence: Alinea, 2000); and Maria Tinacci Mossello, *L'organizzazione del territorio pratese durante la seconda guerra mondiale e la ricostruzione* (Florence: Università degli Studi, 1983).

10. Manufacturing industries operated chiefly in the following sectors: mechanics and metals (Florence, Scandicci, Sesto Fiorentino, Calenzano), fashion and clothing (Florence, Campi Bisenzio, Sesto Fiorentino, Calenzano, Signa, Lastra a Signa), chemistry (Florence, Calenzano, Sesto Fiorentino, Scandicci), textiles (the whole Prato area, Campi Bisenzio, Calenzano, Signa, Lastra a Signa), wood and furniture (Florence, Scandicci, Sesto Fiorentino, Bagno a Ripoli, Lastra a Signa), printing and publishing (Florence, Scandicci, Sesto Fiorentino), nonmetallic minerals (Sesto Fiorentino, Florence, Lastra a Signa, Campi Bisenzio, Impruneta), leather (Florence, Scandicci, Signa, Lastra a Signa), and food and tobacco (Florence, Scandicci, Sesto Fiorentino, Calenzano, Campi Bisenzio, Lastra a Signa, Fiesole).

11. ISTAT, *Censimento generale della popolazione. Risultati provinciali e comunali sulla popolazione e sulle abitazioni* (Rome: Istat, 1982).

12. The most polluted tributaries of the Arno were the Bisenzio, since it received the wastewater of the textile industries located in Prato, Calenzano, Sesto Fiorentino, Campi Bisenzio, and Signa; the Vingone, which crossed two of the most densely industrial areas of the province (Scandicci and Lastra a Signa); the Ombrone, whose waters contained high concentrations of cleansing agents, sulfur, ammonia, phos-

phates, sulfates, chlorines, and cyanides; and the Elsa, contaminated with phosphates, cleansing agents, chlorides, chromium, and cyanides.

13. The cost of $1m^3$ of self-produced water (from wells) amounted to 10 or 20 Italian liras; that of 1 m^3 bought from the municipal utilities varied from 45 to 90 liras.

14. Law no. 615, 1966, "Provvedimenti contro l'inquinamento atmosferico." This is the first law approved in Italy that contains rules to prevent air pollution. The law divided the national territory into two control zones, called A and B. Zone A included the municipalities of central-northern Italy with a population from 70,000 to 300,000 inhabitants and those with a numerically lower population but with industrial, urban, geographical and meteorological characteristics particularly unfavorable with regard to atmospheric pollution; the municipalities of southern and insular Italy with population from 300,000 inhabitants to 1 million and those with a numerically lower population but with industrial, urban, geographical, and meteorological characteristics particularly unfavorable with regard to atmospheric pollution; the municipalities that do not fall into the two previous categories, but that had particular problems regarding air pollution.

15. Comune di Firenze, *L'inquinamento delle acque: Atti ufficiali del convegno* (Florence: Comune di Firenze, Jan. 29, 1971).

16. Law No. 125, Mar. 3, 1971. The measure prohibited the production, trade, importation, and use of nonbiodegradable detergents.

17. "Proposti valori standard unici per gli effluenti domestici e industriali," *Toscana Consiglio Regionale* 2 (1972): 53–55.

18. Regional Law No. 22, May 27, 1974, *Interventi per il reperimento e l'utilizzazione di risorse idriche, nonché per lo smaltimento e la depurazione delle acque di rifiuto.*

19. "Una nuova politica del suolo e delle acque per lo sviluppo economico-sociale del bacino dell'Arno," *La Regione*, Nov. 1970.

20. "Osservazioni su Le linee del programma regionale di sviluppo economico," *Toscana Domani*, Nov.–Dec. 1973.

21. "Approvato lo schema di progetto pilota per la sistemazione del bacino dell'Arno," *Toscana Consiglio Regionale* 3 (1973): 55–60.

22. Regione Toscana, *Progetto pilota per la sistemazione del bacino dell'Arno: Rapporto finale*, vol. 2, *Organizzazione del progetto*, 140–42 (Florence: Regione Toscana, 1978).

23. Regione Toscana, *Piano regionale di risanamento delle acque: Primo programma di intervento 1980–1982* (Florence: Regione Toscana, 1980).

24. Comune di Firenze, *Firenzecologia* (Rome: Il Ventaglio, 1987), 74–85.

25. Comune di Firenze, *Firenzecologia*, 74–85.

26. The plant produced 300–500 kg per day of chloridic acid, 2–4 kg of heavy metals (lead, zinc, mercury, and cadmium), and 20–40 g of organic chlorine derivatives, of which 5 g was dioxin. Comune di Firenze, *Firenzecologia*, 93.

27. "L'inquinamento dell'Arno: Il risanamento è un problema di priorità nazionale e regionale; Approvata una mozione per la realizzazione degli obiettivi previsti dalla legge 319/76 e dal piano regionale di risanamento," *Toscana Consiglio Regionale* 14 (1984): 81.

28. "Integrata la disciplina urbanistica per una migliore pianificazione territoriale," *Toscana Consiglio Regionale* 14 (1984): 521.

29. "L'energia in Toscana: Problemi e prospettive," *Toscana Consiglio Regionale* 16 (1986):, 377.

30. Cgil Regionale Toscana, "Seminario sui problemi energetici," *Impruneta*, Sept. 9–10, 1986.

31. The function of the reservoir was to stabilize the flow of the Arno in a way that aided in the capture of water and other functions of purification plants.

CHAPTER 3: FLORENCE AND ITS WATERS

1. This chapter is based on secondary sources—the books containing reports on studies and researches published by the Commune of Florence (Comune di Firenze), by the Regional Administration of Tuscany (Regione Toscana), and by the Arno River Basin Authority (Autorità di Bacino del Fiume Arno)—and primary sources such as the Official Records of Florence City Council Proceedings (hereafter ACCF), the Edoardo Detti Records in the State Archive of Florence (hereafter ASF-EDR), and the Archives of the Region of Tuscany (hereafter ART). The archive documents have been translated into English in order to make them accessible to an international audience.

2. Federico Paolini, *Firenze 1946–2005: Una storia urbana e ambientale* (Milan: FrancoAngeli, 2014).

3. See Donald Worster, *Rivers of Empire: Water, Aridity and the American West* (Oxford: Oxford University Press, 1985); Theodore Steinberg, *Nature Incorporated: Industrialization and the Waters of New England* (Cambridge: Cambridge University Press, 1991); John Opie, *Ogallala: Water for a Dry Land* (Lincoln: University of Nebraska Press, 2000); Martin V. Melosi, *Effluent America: Cities, Industry, Energy and the Environment* (Pittsburgh: University of Pittsburgh Press, 2001); Mark Cioc, *The Rhine: An Eco-Biography, 1815–2000* (Seattle: University of Washington Press, 2002); Petri S. Juuti and Tapio S. Katko (eds.), *From a Few to All: Long-term Development of Water and Environmental Services in Finland* (Pieksämäki, Finland: KehräMedia, 2004); Harold L. Platt, *Shock Cities: The Environmental Transformation and Reform of Manchester and Chicago* (Chicago: University of Chicago Press, 2005); Jamie Benidickson, *The Culture of Flushing: A Social and Legal History of Sewage* (Vancouver: UBC Press, 2007); Joseph W. Dellapenna and Joyeeta Gupta, *The Evolution of the Law and Politics of Water* (Dordrecht: Springer Netherlands, 2009); Carl Smith, *City Water, City Life: Water and the Infrastructure of Ideas in Urbanizing Philadelphia, Boston, and Chicago* (Chicago: University of Chicago Press, 2013).

4. Ugo Losacco, "Notizie e considerazioni sulle inondazioni d'Arno in Firenze," *L'Universo*, no. 5 (1967): 720–820.

5. Mariella Zoppi, *Firenze e l'urbanistica: La ricerca del piano; La politica amministrativa e i piani urbanistici dalla ricostruzione ad oggi* (Florence: Edizioni delle autonomie, 1982); Lando Bortolotti and Giuseppe De Luca, *Come nasce un'area metropolitana: Firenze, Prato, Pistoia, 1848–2000* (Florence: Alinea, 2000); Simone Neri Serneri, *Incorporare la natura: Storie ambientali del Novecento* (Rome: Carocci, 2005), 219–77.

6. John R. McNeill, *Qualcosa di nuovo sotto il sole: Storia dell'ambiente nel XX secolo* (Turin: Einaudi, 2002), 377–413 (original edition: *Something New under the Sun: An Environmental History of the Twentieth Century World* [New York, W. W. Norton, 2000]).

7. Fabiani administration from November 29, 1946, to July 5, 1951 (left); La Pira administration from july 6, 1951, to august 9, 1956, and august 3, 1956, to june 16, 1957 (Christian Democrat Party); Prefectorial Commissioner Lorenzo Salazar from june 27, 1957, to march 7, 1961; La Pira administration from march 7, 1961 to march 9, 1965 (center-left); Lagorio administration from march 10, 1965, to november 19, 1965 (center-left); Prefectorial Commissioner Adriano Monarca from november 16, 1965, to july 29, 1966.

8. ACCF, Statement by Mayor Fabiani, july 12, 1948: "When the Administration is unable to inscribe in the budget an optional sum of one hundred lire, which would not even be considered in a budget of 1935, you mustn't believe that much can be done!"

9. ACCF, Statement by Councillor Pucci, July 12, 1948: "It is the problem of unemployment the one that more urgently presents itself in the political and economic context. If you took the trouble [addressing Christian Democrat Franchini] to come one day to see the room where we work in Palazzo Vecchio, you would see how daily 40, 50, 60 people come to us . . . one has an ill wife, another has his parents to care for, and every day these put pressure on us for the possibility of earning a loaf of bread."

10. Edoardo Detti has been one of the most important urban planners in Italy, director of the City Planning Institute of the University of Florence and chairman of the Town Planning National Institute (INU, Istituto Nazionale di Urbanistica).

11. Letter by Detti to the Socialist Senator Luigi Mariotti, Feb. 17, 1961, Envelope 9, Document 1133, ASF-EDR: "Prof. La Pira, has, as always, made various detours at this time on that question, and now he says that the Commune cannot confront it. The question concerns instead municipal city planning policies. Demagogic weaknesses would not afterward be accepted, neither would I condone them, or enter into arguments"; Letter by Detti to La Pira, Enriques Agnoletti, Giulio Mayer, Jan. 3, 1962, Envelope 9, Document 1934, ASF-EDR.

12. Letter from Domizio Caderni to Detti, [undated but datable to 1961–1962], Envelope 14, Document 1657, ASF-EDR: "I would pray to give permits and to examine zonings since in this case allotment prices would reach frightening levels . . . since with

such block speculation increases every day (and by going around offices I hear that all blame is made to fall upon Architect Detti)."

13. *Observations Concerning the City Plan by the Association of Industrialists of the Province of Florence*, Envelope 19, Document 2094, ASF-EDR.

14. Davis Ottati, *Il ventre di Firenze: Storia della fognatura dall'epoca romana a oggi* (Florence: Editoriale Olimpia, 1999).

15. Davis Ottati, *L'acquedotto di Firenze dal 1860 ad oggi* (Florence: Vallecchi, 1983).

16. Comune di Prato, Assessorato alle Aziende Municipalizzate, *L'approvvigionamento di acqua potabile nel quadro generale del problema idrico di Prato: Relazione a cura del Servizio Acquedotto dell'Azienda servizi municipalizzati*, Convegno di studio sui problemi idrici del comprensorio pratese, Prato, June 13–14, 1964 (Florence, 1964), 14–15.

17. This paragraph is based on documents contained in ART, Office of the Provincial Health Officer of Florence, classified under the entries Waters and Sanitary Measures.

18. Regione Toscana, *Mappa degli inquinamenti idrici della Toscana* (Florence, 1974).

19. Regione Toscana, *Mappa degli inquinamenti idrici*, 62.

20. Regione Toscana, *Mappa degli inquinamenti idrici*, 84.

21. Piero Innocenti, *L'industria nell'area fiorentina* (Florence: Associazione degli industriali della provincia di Firenze, 1979), 702.

22. Regione Toscana, *Mappa degli inquinamenti idrici*, 102–5.

23. Franco Nencini, *Firenze, i giorni del diluvio* (Florence: Sansoni, 1966), 31–32.

24. This was the opinion of two city planners of the University of Florence: "The level of the Arno increases therefore at a disturbing rate and, between three and four o'clock in the morning of the 4th, from the dams of Levane and La Penna began a flow—with sluice gates which, if not completely open, released at any rate all the water that they received—well in access of 2,000 m^3 per second, a flow described as catastrophic by the very construction projects of the dams." See Ilario Principe and Paolo Sica, "L'inondazione di Firenze del 4 novembre 1966," *L'Universo* 47, no. 2 (1967): 194. On June 3, 1966, the councillor for agriculture of the Provincial Administration of Florence, Riccardo Degl'Innocenti, wrote a letter to the ENEL (National Electricity Board), District of Florence, regarding the Levane and La Penna dams: "This Councillorship Office . . . has been repeatedly and forcefully recalled to actively take note of the damages and sudden floods of the Arno, caused by the opening of the hydroelectric power plants of Levane and La Penna to the fish fauna of the river. . . . We do not intend to raise here a grave problem, which does exist and does cause serious situations, of the threat which such floods represent for the safety of the citizens"; Province of Florence, *La Provincia di Firenze e l'Unione regionale delle provincie toscane contro le calamità naturali per una politica di difesa del suolo e di*

riordino idrogeologico: Note e documenti 1956–1966 (Florence, 1966), 7, 21. See also Martin Reuss, "Floods," in Shepard Krech, John R. McNeill, and Carolyn Merchant (eds.), *Encyclopedia of World Environmental History*, vol. 2 (New York: Routledge, 2004), 539.

25. Nencini, *Firenze, i giorni*, 39–42; Giacomo De Angelis, *Le acque dell'Arno* (Lanciano, 1969).

26. Nencini, *Firenze, i giorni*, 36–37.

27. Provincia di Firenze, *La Provincia di Firenze e l'Unione regionale*, 27–28.

28. Two administrations were in office during the biennium of the flood: the Bargellini administration from July 29, 1966, to November 3, 1967 (center-left) and the Bausi administration (center-left) from November 3, 1967, to August 26, 1968.

29. See the report on hydraulic hazard published by the Arno River Basin Authority: "The urban development in areas pertaining to the riverbed, also occupying the overflow areas, has aggravated further the risk of river inundations. Such urban development in areas subject to the risk of flood has occurred mainly after 1967, as is shown by aerial photographs covering the region, and it had abundantly established itself by 1974: and it is still under way"; Autorità di Bacino del Fiume Arno, *Trasformazioni del territorio e sviluppo dell'edificato lungo il corso dell'Arno e degli affluenti (1954–1993 e 1995)* (Florence, 1997), 27.

30. From 1968 to 1980 the following administrations were in office: Bausi administration from August 26, 1968, to April 28, 1969 (center-left); Prefectorial Commissioner Guido Padalino from April 28, 1969, to September 22, 1970; Bausi administration from September 16, 1970, to September 7, 1973, and September 7, 1973, to September 16, 1974 (center-left); Zoli administration from September 4, 1974, to October 30, 1974 (Christian Democrat Party); Prefectorial Commissioner Antonio Lattarulo from October 28, 1974, to July 26, 1975; Gabbugiani administration from July 25, 1975, to September 3, 1980 (left).

31. Comune di Firenze, *L'inquinamento delle acque: Atti ufficiali del convegno* (Florence, 1971).

32. Regione Toscana, Law No. 22, May 27, 1974, "Interventi per il reperimento e l'utilizzazione di risorse idriche, nonché per lo smaltimento e la depurazione delle acque di rifiuto."

33. This law—centered on sanitary rather than environmental regulatory aspects—focused on wastewater and discharge regulations and represented a break with traditional policies where water was considered essentially as a resource to be exploited.

34. Regione Toscana, *Piano regionale di risanamento delle acque: Primo programma di intervento 1980–1982* (Florence, 1980).

35. Commissione Interministeriale per lo Studio della Sistemazione Idraulica e della Difesa del Suolo, *La sistemazione del Bacino dell'Arno: Relazione sul lavoro svolto al 31 dicembre 1968* (Rome, 1968).

36. Regione Toscana, *Progetto pilota per la sistemazione del bacino dell'Arno: Rapporto finale* (Florence, 1978).

37. Comune di Firenze, *Firenzecologia* (Rome: Il Ventaglio, 1987).

38. Statement by Councillor Casule, Nov. 12, 1973, ACCF.

39. Statement by Councillor Conti, June 5, 1981, ACCF; Statement by Councillor Cariglia, June 5, 1981, ACCF: "The first safe data seem to me that the Bilancino Reservoir, if falling within the scheme of optimization of the regime of the Arno, it certainly will not by itself resolve the problems of the safety of the river. This is said by the experts; it is also said in the most advanced proposals, and therefore this work does not serve, it does not contribute, should it remain the only work, to ensure safety from floods, thus the true reason for the existence of the Bilancino Reservoir should be to protect Florence from thirst."

40. Martin V. Melosi, *The Sanitary City: Urban Infrastructure in America from Colonial Times to the Present* (Baltimore: Johns Hopkins University Press, 2000).

41. See John Dickie, John Foot, and Frank M. Snowden (eds.), *Disastro! Disasters in Italy since 1860: Culture, Politics, Society* (New York: Palgrave, 2002).

42. The measure postponed the deadline for the submission of plans for the cleaning up of waters to the March 31, 1981.

CHAPTER 4: INDUSTRIES, TOURISM, AND THE ENVIRONMENT

1. On industrial development, see Giorgio Mori, "Dall'unità alla guerra: Aggregazione e disgregazione di un'area regionale," Domenico Preti, "Tra crisi e dirigismo: L'economia toscana del periodo fascista," Giacomo Becattini, "Riflessioni sullo sviluppo socio-economico della Toscana in questo dopoguerra," all in Giorgio Mori (ed.), *Storia d'Italia: Le Regioni dall'unità a oggi, La Toscana* (Turin: Einaudi, 1986); Istituto Regionale Programmazione Economica della Toscana (IRPET), *Le zonizzazioni della Toscana 1754–1973* (Florence: Tip. Giuntina, 1988); Giuseppe Pozzana, *Livorno e la sua provincia: Materiali per una storia dello sviluppo economico e sociale dall'unità d'Italia ai giorni nostri; Rassegna bibliografico-critica* (Milan: FrancoAngeli, 1987); Rosella Luchetti and Graziella Poli, *La Magona di Piombino: 1944–1970* (Florence: La Nuova Italia, 1982); Giampiero Celati and Leo Gattini, *Sale e pietra* (Pisa: Giardini, 1993); Giampiero Celati and Leo Gattini, *Cinquant'anni di vita* (Livorno: Nuova Fortezza, 1995); Mario Barberini, *Scarlino e il suo territorio nella evoluzione storica della Maremma* (Pisa: Nistri-Lischi, 1985).

2. Gino Guarnieri, *Livorno marinara: Gli sviluppi portuali, la funzione economica, la tecnica commerciale marittima; Con tavole illustrative fuori testo e ricca appendice documentaria* (Livorno: Tip. Benvenuti e Cavaciocchi, 1962); Maurizio Bettini, *Storia del porto di Livorno 1949–1994* (Livorno: Erasmo, 2004).

3. Solvay S.A., *Solvay: 125° anniversario 1863–1888* (Bruxelles: Solvay, 1988); Giorgio Saviane and Mario Cresci, *Voi di Rosignano* (Milan: Solvay, 1988).

4. Pier Luigi Panciatici, *La siderurgia a Piombino: Uno stabilimento e una città* (Piombino: Officina libraria, 1996); Pino Bertelli and Maurizio Moretti (eds.), *Una centrale al veleno: La battaglia del carbone a Piombino, 1968–1988* (Piombino: TraccEdizioni, 1989); Tiziano Arrigoni and Marisa Giachi (eds.), *Identità di una città industriale: Piombino 1860–1992* (Piombino: Comune di Piombino, 1998); Michele Lungonelli, *La Magona d'Italia: Impresa, lavoro e tecnologie in un secolo di siderurgia toscana 1865–1975* (Bologna: il Mulino, 1991).

5. Gianluca Barducci and Giorgio Casule, *Fanghi rossi di Scarlino, dieci anni di esperienze: L'evoluzione produttiva e ambientale dell'unico stabilimento italiano di produzione di biossido di titanio* (Florence: Regione Toscana, 1982); Fabrizio Fuga, *L'industria del biossido di titanio: La tioxide di Scarlino, un case study* (Pisa: Pacini, 1996); Saverio Luzzi, *Il virus del benessere: Ambiente, salute, sviluppo nell'Italia repubblicana* (Roma-Bari: Laterza, 2009), 129–30.

6. On the development of tourism, see Giuliana Mariotti, *Storia del turismo* (Rome: Edizioni Saturna, 1958); Maria Clelia Cadorna, *La storia della villeggiatura: Dall'epoca romana al Novecento* (Rome: Edizioni Abete, 1994); Marc Boyer, *Il turismo: Dal gran tour ai viaggi organizzati*, trans. Silvia Marzocchi (Turin: Universale Electa/Gallimard, 1997); Patrizia Battilani, *Vacanze di pochi, vacanze di tutti: L'evoluzione del turismo europeo* (Bologna: il Mulino, 2001). On tourism in Tuscany, see Liliana Treves (ed.), *Le risorse economiche del turismo in Toscana* (Padua: Cedam, 1967); Confesercenti regionale, *Il turismo in Toscana: Atti del Convegno* (Poggibonsi: Lalli, 1984); IRPET, *Il turismo in Toscana: Rapporto strutturale 1993–2000* (Florence: Regione Toscana, 2001).

7. On the environmental problems discussed in this chapter, see: ARPAT, *Segnali ambientali in Toscana 2001* (Florence: Edifir, 2001); ARPAT, *Segnali ambientali in Toscana 2005* (Florence: Edifir, 2005); Regione Toscana, *Piano Regionale di Azione Ambientale 2004–2006* (Florence: Edifir, 2007); ARPAT, *Relazione sullo stato dell'ambiente in Toscana 2008* (Florence: Edifir, 2008); ARPAT, *Segnali ambientali in Toscana 2009* (Florence: Edifir, 2009); ARPAT, *Relazione sullo stato dell'ambiente in Toscana 2009* (Florence: Edifir, 2009); Marco Mazzoni (ed.), *Rapporto sullo stato delle acque dei principali fiumi in Toscana* (Florence: Centro stampa 2P, 2001); Marco Mazzoni (ed.), *Rapporto sullo stato delle acque marine in Toscana* (Florence: Centro stampa 2P, 2001); ARPAT, *La qualità delle acque marine costiere in Toscana* (Florence: Edifir, 2004).

8. Comune di Piombino, *L'ambiente Piombino: Cinque anni di attività per la tutela ambientale* (Piombino, 1991); Commissione paritetica Comune di Piombino-ENEL per il controllo della centrale termoelettrica di Torre del Sale, *Lo stato della qualità dell'aria* (Pisa: Tipografia editrice pisana, 1992); Marcello Mossa Verre (ed.), *Analisi del rischio per l'area di Piombino e strategie d'intervento: Piano di risanamento per le aree critiche a elevata concentrazione industriale di Livorno e Piombino* (Florence: ARPAT, 2000).

PART II: MOBILITY AND THE ENVIRONMENT

1. For a historiographic picture beyond Italy, see Massimo Moraglio, "Niche Development or Wider Renaissance? Italian Mobility History Growth in the Last Decade," 167–72; Valentina Fava, "Production versus Mobility? New Perspectives for an Old Dilemma," 247–52; Tom McCarthy, "A Natural Intersection: A Survey of Historical Work on Mobility and the Environment," 61–81; all in Gijs Mom, Gordon Pirie, and Laurent Tissot (eds.), *Mobility in History: The State of the Art in the History of Transport, Traffic and Mobility* (Neuchâtel: Éditions Alphil, 2009).

2. An international workshop, "Mobility and the Environment" (June 3–6, 2010), was held at the Rachel Carson Center of the Ludwig-Maximilians Universität Munich with the aim of starting a new series of research expressly devoted to environmental mobility studies. The workshop was directed by Gijs Mom (Eindhoven University of Technology) and Clapperton Mavhunga (Massachusetts Institute of Technology).

3. Federico Paolini, "Automobili e ambiente: Verso una storia ambientale della mobilità?" 5–13; Tom McCarthy, "Automobili, politica ambientale e individualismo liberale: Un dilemma americano," 15–32; Massimo Moraglio, "Una forma goffa e sgraziata: Paesaggio e autostrade in Italia nel secondo dopoguerra," 33–46; Sigfrido Ramírez Pérez, "Un'auto più pulita e sicura: La Comunità europea e l'avvio della standardizzazione per gli autoveicoli (1952–1970)," 47–61; Christopher W. Wells, "La morte del Modello T: Strade pavimentate, auto coperte e tecnologia desueta," 63–75; all in *I Frutti di Demetra* 21 (2009). Some economists may be included within the context of environmental mobility studies. The Italian Society of Transportation Economics and Logistics (Società Italiana di Economia dei Trasporti e della Logistica/Siet) was established in 1986. Some members of this organization have produced interesting studies on the interactions between transportation and the environment: Romeo Danielis, *I trasporti e l'ambiente: Aspetti di economia e politica dei trasporti* (Turin: Giappichelli, 1996); Patrizia Lattarulo (ed.), *I costi ambientali e sociali della mobilità* (Milan: FrancoAngeli, 2003). See also the following working papers: Gerardo Marletto, "Getting Out of the Car: An Institutional/ Evolutionary Approach to Sustainable Transport Policies" (working paper, Crenos, University of Cagliari, 2008); Marletto, "Changing Regime: From the Car to 3Bs (Bus, Bicycle and Batteries)" (working paper, Isfort, Rome, 2011).

4. On the importance of the automobile as a player in social and environmental change, see Raymond Flower, *One Hundred Years on the Road: A Social History of the Automobile* (New York: McGraw Hill, 1981); James J. Flink, *The Automobile Age* (Cambridge, MA: MIT Press, 1988); Karin Sandquist, *The Appeal of Automobiles: Human Desires and the Proliferation of Cars* (Stockholm: Swedish Transport and Communications Research Board, 1997); David Thoms, Tim Claydon, and Len Holden,

The Motor Car and Popular Culture in the 20th Century (Aldershot, UK: Ashgate, 1998); Mark S. Foster, *A Nation on Wheels: The Automobile Culture in America since 1945* (Belmont, UK: Thomson, 2003); Federico Paolini, *Un paese a quattro ruote: Automobili e società in Italia* (Venice: Marsilio, 2005); Tom McCarthy, *Auto Mania: Cars, Consumers and the Environment* (New Haven, CT: Yale University Press, 2007); Brian Ladd, *Autophobia: Love and Hate in the Automotive Age* (Chicago: University of Chicago Press, 2008); Christopher W. Wells, *Car Country: An Environmental History* (Seattle: University of Washington Press, 2013); Gijs Mom, *Atlantic Automobilism: Emergence and Persistence of the Car 1895–1940* (New York: Berghahn, 2014).

5. Automobile Club d'Italia, *XXX Conferenza del traffico e della circolazione* (Rome, 1973).

6. See Laura Scichilone, *L'Europa e la sfida ecologica: Storia della politica ambientale europea (1969–1998)* (Bologna: il Mulino, 2008).

CHAPTER 5: THE ENVIRONMENTAL IMPACT OF THE CAR ON ITALIAN CITIES, 1950–1974

1. Cities of the center-north (in order of size): Milan, Brescia, Turin, Genoa, Florence, Bologna, Venice, Padua, Trieste, Livorno-Pisa, Verona, La Spezia–Carrara, Ferrara, Biella, Parma-Modena–Reggio Emilia, Rimini-Ancona, Udine, Vicenza, Alessandria, Bassano del Grappa, Piacenza, Trento, Bolzano; of the center-south: Naples, Rome, Palermo, Catania, Bari, Messina, Taranto, Reggio Calabria, Cagliari, Pescara, Sassari, Terni.

2. Tecneco, *Prima relazione sulla situazione ambientale del paese* (Rome, 1973), 2:284–400. In particular, the area of Milan-Brescia went from 3,803,992 residents to 6,351,017 and Turin from 834,375 to 1714,292; in the center-south, Rome and Naples registered the highest increase from 1,734,055 residents to 3,117,232 and from 2,713,015 to 3,652,745, respectively.

3. To expand on the subject of events in Italian city planning, see Giuseppe Campos Venuti and Federico Oliva (eds.), *Cinquant'anni di urbanistica in Italia 1942–1992* (Roma-Bari: Laterza, 1993); Francesco Indovina (ed.), *Lo spreco edilizio* (Padua: Marsilio, 1972); Cesare De Seta, *Città, territorio e Mezzogiorno in Italia* (Turin: Einaudi, 1977).

4. Giuseppe Dematteis, "Le trasformazioni territoriali e ambientali," in *Storia dell'Italia repubblicana*, vol. 2, *La trasformazione dell'Italia: Sviluppo e squilibri* (Turin: Einaudi, 1995), 666.

5. Bernardo Secchi, "Il settore edilizio e fondiario in un processo di sviluppo economico," in Indovina, *Lo spreco*, 3–46.

6. Giuseppe Campos Venuti, "Cinquant'anni: Tre generazioni urbanistiche," in Campos Venuti and Oliva, *Cinquant'anni*, 13–34.

7. Campos Venuti, "Cinquant'anni," 15.

8. The data have been calculated by the author from statistics provided by Automobile Club of Italy (ACI), *Autoveicoli circolanti*, 1950–1975; and ISTAT, *Annuario statistico italiano*, 1950–1975.

9. "L'Italia con l'auto alla gola," *La Domenica del Corriere*, Oct. 15, 1968.

10. See *Notiziario Anci*, 1958–1974, especially issues 12 (1962), 6 (1965), 10–11 (1965), 7–8 (1967), 10–11 (1967).

11. The Automobile Club of Italy was founded on January 23, 1905. The establishment of an association of drivers was felt necessary after 1898, when an earlier automobile club had been founded in Turin, followed by others in Florence, Milan, and Genoa. In 1960 the Automobile Club of Italy had 305,500 members. In 1975 ACI was officially acknowledged as a "public institution."

12. The law of 1942 was restricted to prescribing that general town planning programs should indicate a network of main roads.

13. ANCI was created in 1901 as a free league of communes. Until 1926, when the Fascist government included it in a confederation of autarchic institutions, it carried out propaganda activities concerning local autonomy, causing an interesting meeting point between Catholic, socialist, and liberal-democrat administrators. In the postwar period ANCI regained its original role and independence.

14. "Mozione degli amministratori di comuni e province presenti alla XXIV Conferenza di Stresa," *Notiziario Anci*, Oct.-Nov. 1967.

15. Born in 1916, the Lega dei Comuni Socialisti (League of Socialist Communes) was suppressed in 1926 by the Fascist regime. During the postwar period it changed its name to Lega dei Comuni Democratici, delle Regioni, delle Province e degli Enti Minori (League of Democratic Communes, Regions, Provinces, and Minor Institutions) following its joining the Republicans (in Umbria) and the Communist Party. At the Fifth Congress of Bologna in 1968, it changed its name into Lega per l'Autonomia dei Poteri Locali (League for the Autonomy of Local Authorities).

16. "Sul problema del traffico," *Il Comune Democratico*, Mar. 1967.

17. See Filippo Ciccone, "Roma: Capitale senza piano," in Campos Venuti and Oliva, *Cinquant'anni*, 241–58.

18. On the events concerning the urban disfigurement of Rome, see Antonio Cederna, *I vandali in casa* (Roma-Bari: Laterza, 1956); Cederna, *Mirabilia urbis: Cronache romane 1957–1965* (Turin: Einaudi, 1965).

19. "Un'automobile ogni metro quadro," *Quattroruote*, Feb. 1956.

20. "Roma si ferma per il caos automobilistico: Un record mondiale; 5 chilometri all'ora," *L'Espresso*, Nov. 6, 1966.

21. Commissione per l'Indagine sul Traffico a Roma, *Il traffico a Roma: La situazione attuale e le previsioni fino al 1985* (Rome: Amministrazione comunale, 1968).

22. "Commercianti contro urbanisti per l'isola di Roma," *L'Espresso*, Jan. 9, 1966.

23. "L'inquinamento atmosferico e lo sviluppo del traffico automobilistico," *Strade e traffico*, Nov. 1968.

24. Libero Carriero, *Il costo di eliminazione dell'inquinamento in Italia* (Milan: FrancoAngeli, 1972).

25. On the events concerning the urban growth of Milan, see Federico Oliva, "Milano: La 'deregulation' urbanistica," in Campos Venuti and Oliva, *Cinquant'anni*, 223–40; "Politica urbanistica e sviluppo di Milano dal 1968 ad oggi," *Il Comune Democratico*, June 1974; "Un'alternativa urbanistica per Milano," *Il Comune Democratico*, Sept. 1974.

26. "Il verde in città," *Quattroruote*, June 1961; see also Antonio Cederna, *La distruzione* (Turin: Einaudi, 1975), 324–37.

27. "I gas nocivi emessi dagli autoveicoli," *Inquinamento*, June 1971.

28. Carriero, *Il costo di eliminazione dell'inquinamento in Italia*, 40–50.

29. For a comprehensive coverage of urban developments in Florence, see Giuseppe Campos Venuti and Odoardo Reali, "Firenze: L'urbanistica contrattata," in Campos Venuti and Oliva, *Cinquant'anni*, 313–27.

30. "Otto giorni a piedi," *L'Espresso*, Jan. 9, 1966.

31. "Contributo alla conoscenza della composizione chimica del pulviscolo atmosferico campionato nell'area urbana di Firenze," *Inquinamento*, July-Aug. 1973.

32. Tecneco, *Prima relazione*, 284–400.

33. Senato della Repubblica, *Problemi dell'ecologia*, 2 vols. (Rome, 1971).

34. "Il verde in città," *Quattroruote*, June 1961. A comparison with others European cities: Amsterdam 30 m^2 per person, Stockholm 25, Cologne 20, Munich 16, Copenhagen 12, Moscow 11, London 11, Paris 8.

35. Provinces with high-risk areas: Trieste, Gorizia, Venice, Padua, Turin, Genoa, La Spezia, Livorno, Siracusa, Caltanissetta, Bari, and Brindisi. Metropolitan areas: Milan (including Varese, Como, and Bergamo), Rome, Naples.

36. Renato Squillante, "Traffico e ambiente," in ACI, *XXX Conferenza del traffico e della circolazione di Stresa* (Rome, 1973).

37. "La malaria delle città," *Pirelli*, Jan.-Feb. 1958.

38. "L'inquinamento atmosferico nei riguardi dell'incidente stradale," *Automobilismo e automobilismo industriale*, Jan.-Feb. 1961.

39. "Il traffico automobilistico e l'inquinamento atmosferico," *Le strade*, Mar. 1962; "Aria pericolosa" and "Il veleno nel serbatoio," *l'automobile* 31 (1962); "Non c'è legge contro il gas," *l'automobile* 33–34 (1962).

40. Ministero della Sanità, *Annali della sanità pubblica* 35 (Sept.-Oct. 1964).

41. Ministero della Sanità, *Annali della sanità pubblica*.

42. Law No. 615, July 13, 1966. The regulations enforcing the law were put into

practice only in 1970 and concerned central heating plants and vehicles with diesel engines; industrial plants and vehicles with gasoline engines (the great majority) were excluded.

43. "L'inquinamento atmosferico e lo sviluppo del traffico automobilistico," *Strade e traffico*, Nov. 1968.

44. Italia Nostra, Ente Provinciale per il Turismo, *Inquinamento atmosferico: Processo all'automobile* (Milan: Grafiche A. Nava, 1970).

45. Law No. 437, June 3, 1971.

CHAPTER 6: TRANSPORTATION AND THE ENVIRONMENT IN ITALY, 1944–2006

1. On the history of transportation in Italy, see Andrea Giuntini, "Nascita, sviluppo e tracollo delle rete infrastrutturale," in Franco Amatori, Duccio Bigazzi, Renato Giannetti, and Luciano Segreto (eds.), *Storia d'Italia*, vol. 15, *L'industria* (Turin: Einaudi, 1999), 551–616; Andrea Giuntini and Claudio Pavese (eds.), *Reti, mobilità, trasporti: Il sistema italiano tra prospettiva storica e innovazione* (Milan: FrancoAngeli, 2004); Stefano Maggi, *Storia dei trasporti in Italia* (Bologna: il Mulino, 2005); Federico Paolini, *Un paese a quattro ruote: Automobili e società in Italia* (Venice: Marsilio, 2005); Massimo Moraglio, "Niche Development or Wider Renaissance? Italian Mobility History Growth in the Last Decade," in Gijs Mom, Gordon Pirie, and Laurent Tissot (eds.), *Mobility in History: The State of the Art in the History of Transport, Traffic and Mobility* (Neuchâtel: Editions Alphil-Presses universitaires suisses, 2009), 167–72.

2. The data used in the essay are taken from: ISTAT, *Statistiche ambientali* (1991, 1993, 1998, 2002, 2009); ISTAT, *Flussi di materia dell'economia italiana: Conti anno 1997; Indicatori anni 1980–2001* (Rome, 2006); European Environmental Agency, *Urban Sprawl in Europe: The Ignored Challenge* (Copenhagen, 2006); ENEA, *Rapporto energia e ambiente 2005: I dati* (Rome, 2005); European Commission, Directorate General for Energy and Transport, *Energy and Transport in Figures* (Bruxelles, 2003); Eurostat, *Transport and Environment: Data 1980–1999* (Luxembourg, 2001).

3. The data for municipal roads in 1980 and 1990 relate only to suburban roads. The administrative organization of the road system has been substantially altered after the reformation of Title V of the Constitution (2001), which passed part of the state road system to regional management.

4. In 2001 among the fifteen countries taken into consideration, Portugal and Greece occupied, respectively, the penultimate and last places in per capita GDP. In Italy the absence of adequate public transportation networks in the greater urban areas still makes motorcycles the means of transportation with the highest approval rating (an appreciation index of 8.55 compared to 7.90 for cars, 7.43 for the metro-

politan railway, and 6.03 for bus and tramway). See ISFORT, *Ottavo rapporto sulla mobilità urbana in Italia* (Rome, 2011), 22.

5. Historians have begun only in recent years—and not only in Italy—to address in detail the environmental problems caused by transportation. In recent years we have seen an attempt at creating a new interdisciplinary research sector (environmental mobility studies) to bring together all concepts about transportation, the environment, and social history, with the aim of studying from the environmental point of view the many topics concerning mobility. For a review on the state of the discipline, see Tom McCarthy, "A Natural Intersection: A Survey of Historical Work on Mobility and the Environment," in Mom, Pirie, and Tissot, *Mobility in History*, 61–81; Federico Paolini, "Greening Mobility History in Italy: Toward an Interdisciplinary Way to Environmental Mobility Studies?," in Peter Norton, Gijs Mom, Liz Millward, and Mathieu Flonneau (eds.), *Mobility in History: Reviews and Reflections* (Neuchâtel: Éditions Alphil-Presses universitaires suisses, 2011), 145–50.

6. Think of the impacts, from the planning point of view, of the construction of the Fiat (Lingotto, Mirafiori) and Lancia (Chivasso) plants on Turin, or the Alfa Romeo (Portello, Arese) and Innocenti plants on Milan, or the Ansaldo plant on Genoa.

7. The total surface translates to 163.02 km^2. Data regarding roads and railways represent a prudent estimate calculated by multiplying the total length of the network for a width of 7 m (roads), 7 m (double-track railway lines), and 3.5 m (single-track railway lines). Figures on airports and ports come from Ministero delle infrastrutture e dei trasporti, *Conto nazionale delle infrastrutture e dei trasporti anni 2008–2009* (Rome, 2010).

8. Recently the construction of a high-speed rail line from Bologna to Florence has fueled bitter arguments because of damages caused by rail yards to the hydrogeological system of the Mugello Valley (eighty-one streams, thirty-seven springs, thirty wells, and five rural aqueducts dried up and disappeared).

9. Moraglio, "Niche Development or Wider Renaissance?," 167–72.

10. For 1997 ISTAT (see note 2) calculated an input in materials amounting to 2,696,475,579 tons: 22.82% internal extraction of materials (e.g., biomass, fossil fuels, nonenergy minerals), 10.73% importation (raw materials and semi-worked products, finished products), 25.09% auxiliary inputs (e.g., oxygen for combustion and nitrogen for combustion emissions), 2.98% unutilized materials of internal extraction, 38.38% indirect flows associated with imports. The output of materials amounted to 1,932,982,538 tons: 27.73% emissions and solid waste, 7.46% dissipative uses of products and dissipative losses, 5.74% exports (e.g., biomass, fossil fuels, minerals, composite products), 24.13% net growth of stocks (e.g., transportation infrastructure and buildings, machinery, and other durable goods), 17.10% auxiliary outputs (e.g., steam from combustion, human and animal exhalations), 17.84% indirect flows associated with exports.

11. External economies (externalities) are the influences exercised by economic activities, outside market transactions, on production or on the well-being of a community in a positive (external economies) or negative (external diseconomies) way.

12. The series of environmental statistics published by ISTAT (see note 2) contains little information on the environmental impact of different production sectors as the information is generally presented in aggregated form. The *National Account of Infrastructures and Transport* (Rome: Istituto Poligrafico e Zecca dello Stato, 2002), published by the Ministry of Transportation, contains a chapter devoted to external costs, but considers only a few emissions (carbon dioxide, nitrogen oxides, and volatile organic compounds) and the collection of polluting materials.

13. The resident population equivalent is a measure used to express the impact of civil and productive users on the load of a sewage treatment plant. The equivalent population of the industrial sectors is calculated using specific conversion factors which take into account the production process of the raw materials employed: this is, however, a partial measurement since it essentially concerns only the organic substances present in industrial waste. This is nevertheless a useful measure for comparing the pollution load of different production activities.

14. The 2002 edition of ISTAT, *Statistiche ambientali*, does not show the data for various activities, but only the conversion factors, thus it is impossible to estimate the percentage of the total. Comparative data provides the higher conversion coefficients: sugar industry, alcohol, and tobacco 348; coking plants 96; basic food industries 84; chemical industries 68; nuclear fuels industry 65; paper industry, printing, and publishing 60. The conversion factor for the transportation industry is 1.7.

15. ISTAT, *Statistiche ambientali*, 1993.

16. ISTAT, *Statistiche ambientali*, 1993.

17. The concept of virtual water has been developed by the British geographer John Anthony Allan. See John A. Allan, "'Virtual Water': An Essential Element in Stabilizing the Political Economies of the Middle East," *Yale University Forestry & Environmental Studies Bulletin*, no. 103 (1998): 141–49.

18. By admission of the concept's creators, it is very difficult to calculate with precision the volume of virtual water employed in the production of industrial goods. The global average has been estimated as 80 liters/U.S. dollar. Estimates for cars vary from 50,000 liters per unit to 400,000 liters per 1.1 tons. Assuming an average cost of US$10,000 and a volume of 80 liters, virtual water should amount to 800,000 liters. Here I have assumed a conservative value of 150,000 liters/unit produced; for my purpose this suffices to show the existing relationship between production volumes and the volumes of utilized water, and not to calculate plausible volumes of virtual water. For estimates on virtual water present in some products, see www.waterfootprint.org.

19. The estimate for land occupancy has been obtained by conservatively multiplying the number of cars in circulation by 6.3744 m^2 (the surface occupied by a Fiat Punto).

20. In the 1990s concentrations of nitrogen dioxide far exceeded the accepted European Union limit (an annual average of 40 $\mu g/m^3$) in some sample cities such as Milan (86 $\mu g/m^3$), Athens (76), Marseilles (72), Madrid (66), Lyon (64), Paris (64), Monaco (52), Frankfurt (49), and Copenhagen (48). Data for fine particulate matter (PM_{10}) may not be comparable among member states, due to different measurement techniques (the average of all cities per country): Austria (29.6 $\mu g/m^3$), Belgium (23.7), Denmark (14.9), Finland (20.7), France (20.5), Germany (38.8), Greece (34.9), Ireland (27.8), Italy (18.7), Luxembourg (18.3), Netherlands (19.6), Portugal (19.0), United Kingdom (16.1), Spain (19.5), Sweden (8.5), and EU-15 as a whole (22.0).

21. WHO (World Health Organization), *Health Impact Assessment of Air Pollution in the Eight Major Italian Cities* (Bonn: WHO European Centre for Environment and Health, 2002).

22. Fuel consumption is a conservative estimate obtained by considering a distance equal to 3,842 km/year and an average consumption of 1 liter per 15 km. PM_{10} emissions are calculated assuming a production within the urban cycle of 0.1106 g/vehicle-km for the years 1950–1985 and of 0.0946 g/vehicle-km for the years 1995–2001 (with an urban circulation of 3,842 km/year). The energy expenditure is expressed in Megajoule per passenger for 1 km of traveling and relates to a small/medium-capacity car (the number of cars multiplied by 1.5 Mjoule; e.g., the consumption of an SUV is estimated at 7 Mjoule).

23. "Ultimately, from the 12th report contradictions and antinomies emerge that help to understand the complexity of the relationship between Italian citizens and the use of the car: 1) Notwithstanding that the car induces [drivers] to modify daily schedules ever more often according to the demands of heavy traffic, it is used more and more frequently. A possible explanation rests in the disillusion of the car user for not having found an alternative to the private vehicle; 2) The car is used in order to rationalize daily routines. In fact, if four-wheel mobility doesn't grant the certainty of times, it determines an upset of the daily routine, making the daily schedule ever more unpredictable; 3) The car represents a means of protection from and an interface with the outside world. The subject who stays inside the car has different reactions when he is outside, yet the four wheels become ever more a medium for connection between individuals and places, a kind social and relational city network mesh. 4) The car is increasingly a means of mass transport, but the rules for its use (the highway code) are known only to an elite of car users"; ACI-CENSIS, *XIII Rapporto Automobile ACI-Censis* (2004), 5.

CHAPTER 7: CARS AND EMISSIONS REGULATION IN ITALY, 1950–2008

1. As a matter of fact, some studies on pollution had appeared as long ago as the 1920s in the United States. See George A. Burrell, *Vitiation of Garage Air by Automobile Exhaust Gases* (Washington, DC: Bureau of Mines, 1919); Arno C. Fieldner, *Automobile Exhaust Gases in Vehicular Tunnels* (Washington, DC: Bureau of Mines, 1920); Royd R. Sayers, *Effect of Repeated Daily Exposure of Several Hours to Small Amounts of Automobile Exhaust Gas* (Washington, DC: Bureau of Mines, 1929). See also William L. Faith, *Automobile Exhaust and Smog Formation* (San Marino, CA: Air Pollution Foundation, 1957); John C. Esposito and Larry J. Silverman, *Vanishing air: The Ralph Nader study group report on air pollution* (New York: Grossman Publishers, 1970); Donald N. Dewees, *Automobile Air Pollution: An Economic Analysis* (Cambridge, MA: Harvard University, 1971); David Harrison, *Who Pays for Clean Air: The Cost and Benefit of Federal Automobile Emission Controls* (Cambridge, MA: Ballinger, 1975).

2. Atti Parlamentari, I Legislatura, Camera dei Deputati, *Discussioni, Interrogazione dell'on: Palenzona e risposta del ministro dell'Industria e del commercio Campilli*, Dec. 12, 1952.

3. Car density is the measure of the level of motorization used in the statistics published by the Italian Association of the Automotive Industry. It indicates the ratio of cars per resident: the higher the number of residents, the lower the level of development of private motorization.

4. Associazione nazionale fra industrie automobilistiche e affini (ANFIA), *L'automobile in cifre* (Turin, 1950, 1951, 1952).

5. "La malaria delle città," *Pirelli*, Jan.-Feb. 1958.

6. Indeed, a bill was introduced by the Communist Party senator Scotti on November 20, 1958, decreeing that industrial cities or cities with heavy traffic should provide pollution monitoring weather stations. It also authorized communes to take the necessary steps to reduce car traffic whenever the level of air pollution had reached concentrations deemed dangerous. The Scotti bill remained within the Senate Standing Committee on Health and was never passed. Beginning in 1960, pollution caused by motor vehicles was featured in the car press, which provided space devoted primarily to the relationship between the inhalation of poisonous substances in car exhaust and the onset of respiratory diseases. In its first issue of 1961, the journal of the Commission for Industrial Motoring (Commissione per l'Automobilismo Industriale) of the ACI emphasized the role of motor vehicles as sources of air pollution and listed the harmful effects of smog. The following year, the ACI weekly publication highlighted the fact that in Italy there was no law against gas, while the monthly publication of the Touring Club's Experimental Institute for the Roads Commission (Istituto Sperimentale della Commissione Strade) published an article in a preoccupied tone indicating morbidity related to air pollution. See "L'inquinamento atmosferico

nei riguardi dell'incidente stradale," *Automobilismo e automobilismo industriale*, Jan.-Feb. 1961; "Non c'è legge contro il gas," *l'Automobile* 33–34 (1962); "Il traffico automobilistico e l'inquinamento atmosferico," *Le strade*, Mar. 1962.

7. Ministero della Sanità, *Annali della sanità pubblica* 25 (Sept.-Oct. 1964).

8. On parliamentary debate about Law 615, see Atti parlamentari, IV Legislatura, Senato, *XI Commissione (Igiene e sanità)*, sessions Sept. 29, 1965, Oct. 13, 1965, and Oct. 14, 1965; Atti parlamentari, IV Legislatura, *XIV Commissione (Igiene e sanità)*, sessions Apr. 21, 1966, Apr. 28, 1966, and May 5, 1966.

9. See ACI, *XXX Conferenza del traffico e della circolazione: Relazioni e comunicazioni; Stresa, 27–30 settembre 1973* (Rome, 1973); Luciano Fabbri, *La città inquinata* (Florence: Medicea, 1972); Ugo Leone, *L'Italia inquinata* (Naples: Esi, 1970). On legislation in other countries, see Istituto per la Documentazione e gli Studi Legislativi, *Inquinamenti dell'aria, dell'acqua e del suolo* (Milan: Giuffrè, 1971), 45–73; Tom McCarthy, *Auto Mania: Cars, Consumers and Environment* (London: Yale University Press, 2007), 164–75, 178–92; Brian Ladd, *Autophobia: Love and Hate in the Automotive Age* (Chicago: University of Chicago Press, 2008), 139–73; Frank Uekoetter, *The Age of Smoke: Environmental Policy in Germany and the United States* (Pittsburgh: University of Pittsburgh Press, 2009), 202–17, 257–59.

10. Camera dei Deputati, *Indagini conoscitive e documentazioni legislative*, No. 7, *Indagine conoscitiva della XII Commissione permanente: Situazione e prospettive dell'industria automobilistica nazionale* (Rome, 1971).

11. Camera dei Deputati, *Indagine conoscitiva della XII Commissione permanente, Audizione di Giovanni Agnelli and Dante Giacosa*, session Feb. 20, 1969.

12. Camera dei Deputati, *Indagine conoscitiva della XII Commissione permanente, Audizione di Giuseppe Luraghi*, session Feb. 26, 1969.

13. Camera dei Deputati, *Indagine conoscitiva della XII Commissione permanente, Audizione dei dirigenti (Giordana, Masoero, Rossetto, Pecoraro) del Sindacato italiano dell'automobile*, session Apr. 17, 1969.

14. Italia Nostra, Ente Provinciale per il Turismo, *Inquinamento atmosferico: Processo all'automobile* (Milan: Grafiche A. Nava, 1970).

15. Carlo Ripa di Meana was an Italian politician and environmentalist. As a member of the Socialist Party, he has been commissioner for culture and environment in the two European Commissions chaired by Jacques Délors (1985–1992). From 1993 to 1996 he was the national spokesman for the Italian Green Party and from 2005 to 2007 president of Italia Nostra.

16. Italia Nostra, Ente Provinciale per il Turismo, *Inquinamento atmosferico*, 5.

17. "Resoconto stenografico della Tavola rotonda sul regolamento per ridurre l'inquinamento atmosferico causato dai gas di scarico degli autoveicoli," *L'automobile speciale* 6 (1970).

18. "L'inquinamento può essere sconfitto," *L'automobile speciale* 7–8 (1970).

19. Luigi Mammarella, *L'inquinamento in Italia* (Rome: ISVET, 1970); Libero Carriero, *Il costo di eliminazione dell'inquinamento in Italia* (Milan: FrancoAngeli, 1972).

20. "I progressi nella lotta all'inquinamento: Relazione all'assemblea degli azionisti della General Motors (15 marzo 1970)," *Automobilismo e aut, Industriale*, Jan.-Feb. 1970.

21. "Filippo Carpi de Resmini: Che cosa ci riserva il 2000; L'auto distruggerà l'uomo? L'uomo distruggerà l'auto?," *l'Automobile* 20 (1970); "Una politica contro l'inquinamento," *l'Automobile* 43 (1970).

22. On these issues, see Federico Paolini, *Un paese a quattro ruote: Automobili e società in Italia* (Venice: Marsilio, 2005), 111–42; Federico Paolini, *Storia sociale dell'automobile in Italia* (Rome: Carocci, 2007), 63–79; Federico Paolini, "A Country on Four Wheels: The Car and Society in Italy (1900–1974)," *TsT-Transportes, Servicios y Telecomunicaciones* 17 (2009): 108–30.

23. Federico Paolini, "La più grossa sciagura sarebbe se la Fiat entrasse in crisi: Gianni Agnelli, le automobili e l'Italia nei tardi anni Sessanta," *Contemporanea*, no. 1 (Jan. 2004): 85–96.

24. See Paolini, *Un paese*, 223–59.

25. Roland Barthes, *Scritti: Società, testo, comunicazione*, trans. Gianfranco Marrone (Milan: CDE, 1998), 41–50.

26. On these issues, see Ernest Dichter, *Handbook of consumer motivations: the psychology of the world of objects* (New York: McGraw-Hill, 1964); Jean Baudrillard, *La société de consomation: Ses mythes, ses structures* (Paris: Gallimard, 1974); Pierre Bourdieu, *La distinction: Critique sociale du jugement* (Paris: éditions de Minuit, 1979); Baron Isherwood and May Douglas, *World of Goods* (New York: Basic Books, 1979); Chandra Mukerji, *From Graven Images: Patterns of Modern Materialism* (New York: Columbia University Press, 1983); Arjun Appadurai (ed.), *The Social Life of Things: Commodities in Cultural Perspectives* (Cambridge: Cambridge University Press, 1986); Vanni Codeluppi, *I consumatori: Storia, tendenze, modelli* (Milan: FrancoAngeli, 1992); David Thoms, Len Holden, and Tim Claydon, *The Motor Car and Popular Culture in the 20th Century* (Aldershot, UK: Ashgate, 1998); Paolo Capuzzo (ed.), *Genere, generazione e consumi: L'Italia degli anni sessanta* (Rome: Carocci, 2003); Paolini, *Un paese*, 135–87.

27. On the events leading to standardization, see Sigfrido Ramírez Pérez, "Un'auto più pulita e sicura: La Comunità europea e l'avvio della standardizzazione per gli autoveicoli (1952–1970)," *I Frutti di Demetra*, no. 21 (July 2010): 47–61.

28. Atti parlamentari, V Legislatura, Camera dei Deputati, *XIV Commissione (Igiene e sanità)*, sessions Mar. 31, 1971, and Apr. 28, 1971.

29. Tecneco, *Prima relazione sulla situazione ambientale del Paese*, 3 vols. (Rome: Tip. Carlo Colombo, 1973), 1:370–86.

30. Tecneco, *Prima relazione sulla situazione ambientale del paese*, 1:370–86.

31. ACI, *XXX Conferenza del Traffico e della Circolazione: Relazioni e comunicazioni; Stresa, Sept. 27-30, 1973* (Rome, 1973).

32. ACI, *XXX Conferenza del Traffico e della Circolazione.*

33. ACI, *XXX Conferenza del Traffico e della Circolazione.*

34. See Donald N. Dewees, *Economic and Public Policy: The Automobile Pollution Case* (Cambridge, MA: MIT Press, 1974); Frank P. Grad, *The Automobile and the Regulation of Its Impact on the Environment: A Study* (Norman: University of Oklahoma Press, 1975).

35. "La verità sul motore a idrogeno," *Quattroruote*, Jan. 1977; "L'idrogeno non è un'opinione," *Quattroruote*, Mar. 1977; "L'automobile elettrica: come e quando?," *Quattroruote*, Dec. 1977; "Ritorna la benzo elettrica?" *Quattroruote*, July 1977.

36. "Achtung benzene," *Quattroruote*, Sept. 1989.

37. See European Commission, *Everything on Transport Statistics: Data 1970–2002* (dvd-rom) (Luxemburg: European Communities, 2004); OECD, *Environmental Data: Compendium 2004* (Paris: OECD Publishing, 2004).

38. See ISTAT, *Statistiche ambientali: 1996* (Battipaglia: Emilcomp, 1995); ISTAT, *Statistiche ambientali: 1998* (Pomezia: Albagraf, 1998).

39. The targhe alterne is an administrative measure deliberated by the mayors or public authorities for the containment of polluting emissions due to traffic in urban areas. The measure prohibits the use on alternate days of vehicles whose plate has an even or odd number as the last figure. If the authority specifies that on the first day the ban applies to license plates that end with an even number, the second day the ban will start for odd numbered plates, and the third again for even numbered plates, and so on every other day. The measure in this way aims to halve urban traffic.

40. European emissions standards define the acceptable limits for exhaust emissions of vehicles sold in the European Union. The legal framework consists of a series of directives, each amendments to the 1970 Directive 70/220/EEC. The stages are typically referred to as Euro 1, Euro 2, Euro 3, Euro 4, Euro 5, and Euro 6. Since the Euro 2 stage, EU regulations introduced different emission limits for diesel and gasoline vehicles, and for each vehicle type, different standards apply. These are, for example, emission standards regarding carbon monoxide for gasoline passenger cars (g/km): 2.72 (Euro 1); 1.0 (Euro 2); 0.66 (Euro 3); 0.50 (Euro 4, 5, and 6).

41. See Legambiente, *Smog e dintorni 2006: L'inquinamento atmosferico e acustico nelle città italiane* (Rome: Legambiente, 2006); Agenzia per la protezione dell'ambiente, *Qualità dell'ambiente urbano* (Rome: APAT, 2006); Istituto superiore per la protezione e la ricerca ambientale, *Annuario dei dati ambientali 2009* (Rome: ISPRA, 2009).

42. On car and alternative fuel technologies, see Gijs Mom, *The Electric Vehicle: Technology and Expectations in the Automobile Age* (Baltimore: Johns Hopkins

University Press, 2004); John Mikler, *Greening the Car Industry: Varieties of Capitalism and Climate Change* (Cheltenham, UK: Edward Elgar, 2009); Allen E. Fuhs, *Hybrid vehicles and the Future of Personal Transportation* (Boca Raton, FL: CRC Press, 2009); William J. Mitchell, Christopher E. Borroni-Bird, and Lawrence D. Burns, *Reinventing the Automobile: Personal Urban Mobility for the 21st Century* (Cambridge, MA: MIT Press, 2010).

PART III: POLITICS

1. For an overview of Italian history, see John Foot, *Modern Italy* (Basingstoke, UK: Palgrave Macmillan, 2014); Guido Crainz, *Il paese reale: Dall'assassinio di Moro all'Italia di oggi* (Rome: Donzelli, 2012); Paul Ginsborg, *Italy and Its Contents: Family, Civil Society, State, 1980–2001* (London: Penguin Press, 2001). For environmental issues, see Gabriella Corona, *A Short Environmental History of Italy: Variety and Vulnerability* (Winwick, UK: White Horse Press, 2017); Gialuigi Della Valentina, *Storia dell'ambientalismo in Italia: Lo sviluppo insostenibile* (Milan: Bruno Mondadori, 2011).

2. Serge Latouche, "Decrescita," in Giuliano Battiston and Giulio Marcon (eds.), *La sinistra che verrà: Le parole chiave per cambiare* (Rome: Minimum fax, 2018), 34–35.

3. Ernesto Galli della Loggia, "Perché ci manca lo stato," *Corriere della sera*, Aug. 17, 2018.

4. On environmentalism, see Edgar H. Meyer, *I pionieri dell'ambiente: L'avventura del movimento ecologista italiano; Cento anni di storia* (Milan: Carabà, 1995); Andrea Poggio, *Ambientalismo* (Milan: Editrice Bibliografica, 1996); Luigi Piccioni, *Il volto amato della patria: Il primo movimento per la protezione della natura in Italia 1880–1934* (Camerino: Università degli studi di Camerino, 1999); Roberto Della Seta, *La difesa dell'ambiente in Italia: Storia e cultura del movimento ecologista* (Milan: FrancoAngeli, 2000); Gianluigi Della Valentina, *Storia dell'ambientalismo in Italia*. For two highly important books on environmentalism by sociologists, see Mario Diani, *Isole nell'arcipelago: Il movimento ecologista in Italia* (Bologna: il Mulino, 1988); and Donatella Della Porta and Mario Diani, *Movimenti senza protesta? L'ambientalismo in Italia* (Bologna: il Mulino, 2004).

CHAPTER 8: ENVIRONMENTAL KNOWLEDGE, URBAN PLANNING, AND POLITICAL DECISIONS IN ITALY FROM 1850 TO THE PRESENT

1. See Engelbert Schramm, "Experts in the Smelter Smoke Debate," in Peter Brimblecombe and Christian Pfister (eds.), *The Silent Countdown: Essays in European Environmental History* (New York: Springer-Verlag, 1990), 196–209; David Stradling, *Smokestacks and Progressives: Environmentalists, Engineers and Air Quality in America 1881–1951* (Baltimore: Johns Hopkins University Press, 1999); Samuel P. Hays, *Conservation and the Gospel of Efficiency: The Progressive Conservation Movement*

1890–1920 (Cambridge, MA: Harvard University Press, 1999); Clark A. Miller and Paul N. Edwards (eds.), *Changing the Atmosphere: Expert Knowledge and Environmental Governance* (Cambridge, MA: MIT Press, 2001); Barbara L. Allen, *Uneasy Alchemy: Citizens and Experts in Louisiana's Chemical Corridor Disputes* (Cambridge, MA: MIT Press, 2003); Michael Bess, *The Light Green Society: Ecology and Technological Modernity in France 1960–2000* (Chicago: University of Chicago Press, 2003); Stephen Bocking, *Nature's Experts: Science, Politics, and the Environment* (New Brunswick, NJ: Rutgers University Press, 2004); Joseph M. Hodge, *Triumph of the Expert: Agrarian Doctrines of Development and the Legacies of British Colonialism* (Athens: Ohio University Press, 2007); Susan Owens, "Experts and the Environment: The UK Royal Commission on Environmental Pollution 1970–2011," *Journal of Environmental Law* 24 (2012): 1–22; Laurent Coumel, "A Failed Environmental Turn? Khrushchev's Thaw and Nature Protection in Soviet Russia," *Soviet and Post-Soviet Review* 2 (2013): 167–89. For Italy, see Luigi Pellizzoni (ed.), *Conflitti ambientali: Esperti, politica, istituzioni nelle controversie ecologiche* (Bologna: il Mulino, 2011); Piero Bevilacqua, "Scienza e ambiente: Quel che spetta al Novecento," *I Frutti di Demetra* 16 (2008): 5–16. On technological change and the environment, see John R. McNeill, *Qualcosa di nuovo sotto il sole: Storia dell'ambiente nel XX secolo*, trans. Piero Arlorio (Turin: Einaudi, 2000), 377–413.

2. A first reflection is found in Giorgio Nebbia, *Scritti di storia dell'ambiente e dell'ambientalismo 1970–2013*, ed. Luigi Piccioni (Brescia: Fondazione Luigi Micheletti-Valori, 2004), esp. 28–89, 306–83.

3. Jean Paul Deléage, *Storia dell'ecologia: Una scienza dell'uomo e della natura*, trans. Takery Capra (Naples: Cuen, 1991), 78.

4. On these environmental ideas, see Pascal Acot, *Storia dell'ecologia*, trans. Silvana Nesi Sirgiovanni (Rome: Lucarini, 1989); Peter H. Marshall, *Nature's Web: An Exploration of Ecological Thinking* (London: Simon & Schuster, 1992); Donald Worster, *Storia delle idee ecologiche*, trans. Elisa Gunella (Bologna: il Mulino, 1994); Keith Thomas, *L'uomo e la natura: Dallo sfruttamento all'estetica dell'ambiente, 1500–1800*, trans. Elda Negri Monateri (Turin: Einaudi, 1994); Joseph E. De Steiguer, *The Origins of Modern Environmental Thought* (Tucson: University of Arizona Press, 2006).

5. Federico Paolini, *Breve storia dell'ambiente nel Novecento* (Rome: Carocci, 2009), 33–39, 58–69; Luigi Piccioni, "Nazione, patrimonio, paesaggio: Alle origini del moderno ambientalismo in Europa 1865–1914," *Storia e Futuro*, no. 38 (June 2015), http://storiaefuturo.eu/nazione-patrimonio-paesaggio-alle-origini-del-moderno-ambientalismo-in-europa-1865–1914/.

6. On the birth of the protectionist movement in Italy, see Edgar H. Meyer, *I pionieri dell'ambiente* (Milan: Carabà, 1995); Luigi Piccioni, *Il volto amato della Patria: Il primo movimento per la protezione della natura in Italia 1880–1934* (Trento: Temi, 2014); James Sievert, *The Origins of Nature Conservation in Italy* (New York: Peter

Lang, 2000); Luigi Piccioni, *Primo di cordata: Renzo Videsott dal sesto grado alla protezione della natura* (Trento: Temi, 2010).

7. See Piccioni, *Il volto amato*, 1–12, 101–4.

8. See Guido Zucconi, *La città contesa: Dagli ingegneri sanitari agli urbanisti (1885–1942)* (Milan: Jaca Book, 1989); Roberto Parisi, "Verso una città salubre: Lo spazio produttivo a Napoli tra storia e progetto," *Meridiana* 42 (2001): 53–74; Roberto Parisi, "Le forme d'Igea e le persuasioni di Prometeo: Fabbriche, sanatori e città giardino a Napoli," in Gabriella Corona and Simone Neri Serneri (eds.), *Storia e ambiente: Città, risorse e territori nell'Italia contemporanea* (Rome: Carocci, 2007), 123–40; Maria Luisa Ferrari, "Trasformazioni urbane e igiene nell'industrializzazione ottocentesca di Verona, Vicena e Padova," in Corona and Neri Serneri, *Storia e ambiente*, 55–85.

9. In Italy the first legislative provision on hygiene was the Regolamento Napoleonico di Polizia Medica, di Sanità Continentale e di Sanità Marittima of 1806. In 1859 the Kingdom of Sardinia approved a health law, and the Crispi-Pagliani law of December 22, 1888, established the General Directorate of Public Health, headed by a physician-hygienist. The basics of the Crispi-Pagliani law was substantially confirmed in the Testo Unico delle Leggi Sanitarie, approved by Royal Decree No. 1265/1934.

10. Martin V. Melosi, *The Sanitary City: Environmental Services in Urban America from Colonial Times to the Present* (Pittsburgh: University of Pittsburgh Press, 2008), 71–167.

11. Gabriella Corona and Simone Neri Serneri, "Città e ambiente nell'Italia contemporanea: Un'introduzione," in Corona and Neri Serneri, *Storia e ambiente*, 19.

12. Simone Neri Serneri, *Incorporare la natura: Storie ambientali del Novecento* (Rome: Carocci, 2005).

13. Law No. 2258, Mar. 20, 1865; Ministerial Instructions, June 20, 1896; Consolidated Law No. 523, July 25, 1904; Ministerial Decree, Feb. 26, 1927 (unhealthy industries); Royal Decree No. 1265, July 27, 1934 (Arts. 217, 227); Law No. 366, Mar. 20, 1941, Arts. 650 and 674 of the Criminal Code and Art. 844 of the Civil Code.

14. Law No. 185, June 12, 1902; Law No. 364, June 20, 1909; Law No. 688, June 23, 1912; Law No. 778, June 11, 1922; Law No. 1089, June 1, 1939; Law No. 1497, June 29, 1939.

15. Royal Decree No. 1584, Dec. 3, 1922 (Gran Paradiso); Law No. 1511, July 12, 1923 (Abruzzo); Law No. 285, Jan. 25, 1934 (Circeo); Law No. 740, Apr. 24, 1935 (Stelvio).

16. Law No. 1150, Aug. 17, 1942. The problem of regulating urban development had been tackled previously with Law No. 2559, June 25, 1865, and Law No. 2892, Jan. 12, 1885. Edoardo Salzano viewed the 1942 law: "It cannot be called a fascist law. It was the product, at a considerable level, of a bitter clash between diehard defenders of private property and those who wanted to set limits to land utilization rights." See Edoardo Salzano, "L'Istituto nazionale di urbanistica," in Giuseppe Campos Venuti and Federico Oliva (eds.), *Cinquant'anni di urbanistica in Italia 1942–1992* (Roma-Bari: Laterza, 1993), 425.

17. Considering the data reported by Piccioni, in 1905 the members of the Club Alpino Italiano and the Touring Club represented 0.16% of the Italian population. For criticisms of the cited laws, see Luigi Falco, *Natura, monumenti, ambiente: La legislazione di tutela dal 1865 alle regioni* (Turin: Celid, 1979); Neri Serneri, *Incorporare la natura*, 172–86; Piccioni, *Il volto amato della Patria*, 18, 247–75; Campos Venuti and Oliva, *Cinquant'anni di urbanistica*.

18. Salvatore Adorno and Simone Neri Serneri, "Per una storia ambientale delle aree industriali in Italia: Introduzione," in Salvatore Adorno and Simone Neri Serneri (eds.), *Industria, ambiente e territorio* (Bologna: il Mulino, 2009), 13–29. For a recent interpretation of the Italian economic parabola, see Emanuele Felice, *Ascesa e declino: Storia economica d'Italia* (Bologna: il Mulino, 2015).

19. On the origins of Pro Natura, see Meyer, *I pionieri dell'ambiente*, 41–46; and Piccioni, *Primo di cordata*, 231–88. On the Italian branch of the WWF, see Meyer, *I pionieri dell'ambiente*, 161–88. See also Roberto Della Seta, *La difesa dell'ambiente in Italia: Storia e cultura del movimento ecologista* (Milan: FrancoAngeli, 2000); Gianluigi Della Valentina, *Storia dell'ambientalismo in Italia: Lo sviluppo insostenibile* (Milan: Bruno Mondadori, 2011); Nebbia, *Scritti di storia dell'ambiente*.

20. The International Union for the Conservation of Nature was founded in 1948. The president of the MIPN and director of the Gran Paradiso National Park, Renzo Videsott, took part in its establishment. In the 1950s and 1960s several agreements were signed concerning the conservation of internal waterways (Steckborn, 1960; Paris, 1961; Berne, 1963), fishing resources (International Whaling Commission, 1946; Tokyo, 1952; Rio de Janeiro, 1966), and flora and fauna (Rome, 1951; Washington, 1957).

21. Mario Diani, *Isole nell'arcipelago: Il movimento ecologista in Italia* (Bologna: il Mulino, 1988), 50.

22. Paul R. Ehrlich, *The Population Bomb* (New York: Ballantine Books, 1968); Garrett J. Hardin, "The Tragedy of the Commons," *Science* 162, no. 3859 (1968): 1243–48; Ezra J. Mishan, *The Costs of Economic Growth* (Harmondsworth, UK: Penguin, 1969); Herman E. Daly (ed.), *Essays toward a Steady-State Economy* (Cuernavaca: Centre Intercultural de Documentación, 1971); Edward Goldsmith and Robert Allen, *A Blueprint for Survival* (Harmondsworth, UK: Penguin, 1972); Donella H. Meadows and Dennis L. Meadows, *The Limits to Growth: A Report for the Club of Rome's Project on the Predicament of Mankind* (New York: Universe Books, 1972). See also Giorgio Nebbia and Luigi Piccioni, "Un tornante del dibattito italiano sull'ambiente: La ricezione dei Limiti dello sviluppo," *Ricerche Storiche* 41, no. 3 (2011): 519–40.

23. In 1971 in the proceedings of the Senate Steering Committee for Ecological Problems the then-president of the National Research Council, Vincenzo Caglioti, concluded with the statement that the "future of humanity" was "closely linked to the most rational utilization of natural resources, to elimination of the various forms of pollution and to the availability of suitable means for regulating demographic

development"; Vincenzo Caglioti, "L'uomo e l'ambiente," in Senato della Repubblica, *Problemi dell'ecologia*, vol. 1, *Nota introduttiva del presidente Fanfani ed illustrazioni dei professori Caglioti, Marini-Bettolo, Nebbia, Passino, Pavan e Tomaselli* (Rome: Tipografia del Senato, 1971), 2, 32. Published in 1973, the *Prima relazione sulla situazione ambientale del Paese* still remains one of the most important documents on the alteration factors of Italian environmental systems; see Tecneco, *Prima relazione sulla situazione ambientale del Paese*, 3 vols. (Rome: Tip. Carlo Colombo, 1973). On the importance of the *Prima relazione*, see Michele Caputo, "Un progetto per la salvaguardia del patrimonio ambientale italiano," *il Mulino* 56, no. 434 (2007): 1042–49.

24. On Amici della Terra, see CENSIS, *Energia, sviluppo, ambiente* (Milan: Garzanti, 1990), 204–6; David R. Brower, *For Earth's Sake: The Life and Times of David Brower* (Salt Lake City, UT: Gibbs Smith Publisher, 1990). On the genesis of Legambiente, see *Arcipelago verde: La prima guida completa per chi ama la natura*, supplement to *Panorama*, no. 899 (July 5, 1983). In 1983, the four main Italian environmental organizations (WWF, Legambiente, LIPU, and Italia Nostra) had an overall membership of 76,000 (0.13% of the Italian population): 30,000 WWF, 17,000 LIPU, 15,000 Legambiente, 14,000 Italia Nostra; see *Arcipelago verde*, 23–65.

25. Salzano, "L'Istituto nazionale di urbanistica," 424. On Italian town planning, see Patrizia Gabellini, "Figure di urbanisti e programmi di urbanistica," in Campos Venuti and Oliva, *Cinquant'anni di urbanistica*, 440–67; Paolo Scattoni, *L'urbanistica dell'Italia contemporanea: Dall'unità ai giorni nostri* (Milan: Newton & Compton, 2004).

26. On town planning reform projects, see Fiorentino Sullo, *Lo scandalo urbanistico: Storia di un progetto di legge* (Florence: Vallecchi, 1964); Vezio De Lucia, *Se questa è una città: La condizione urbana nell'Italia contemporanea* (Rome: Donzelli, 2006); Salzano, "L'Istituto nazionale di urbanistica," 427–30.

27. See Gianfranco Amendola and Paolo Dell'Anno, *La disciplina dell'inquinamento atmosferico* (Naples: Jovene Editore, 1992); Gianfranco Amendola, *Inquinati e inquinatori* (Rome: La Nuova Italia Scientifica, 1980).

28. Federico Paolini, *Firenze 1946–2005: Una storia urbana e ambientale* (Milan: FrancoAngeli, 2014), 248–73. See also Della Seta, *La difesa dell'ambiente in Italia*; Sara Lorenzini, "Ecologia a parole? L'Italia, l'ambientalismo globale e il rapporto ambiente-sviluppo intorno alla conferenza di Stoccolma," *Contemporanea*, no. 3 (July-Sept. 2016): 395–418.

29. See Sullo, *Lo scandalo urbanistico*; De Lucia, *Se questa è una città*; Salzano, "L'Istituto nazionale di urbanistica"; Paolo Berdini, *Breve storia dell'abuso edilizio in Italia: Dal ventennio fascista al prossimo futuro* (Rome: Donzelli, 2010); Giuseppe Campos Venuti, *Città senza cultura: Intervista sull'urbanistica (a cura di Federico Oliva)* (Roma-Bari: Laterza, 2010).

30. Salzano, "L'Istituto nazionale di urbanistica," 430.

31. Roland Barthes, *Scritti. Società, testo, comunicazione*, trans. Gianfranco Marrone (Milan: CDE, 1998), 41–50.

32. Arno C. Fieldner, *Automobile Exhaust Gases in Vehicular Tunnels* (Washington, DC: Department of the Interior, Bureau of Mines, 1920); American Association for the Advancement of Science, Committee on Science in the Promotion of Human Welfare, Air Conservation Commission, *Air Conservation Report* (Washington, DC, 1965); Ronald G. Ridker, *Economic Costs of Air Pollution: Studies in Measurement* (New York: F. A. Praeger, 1967); Thomas G. Aylesworth, *This Vital Air, This Vital Water: Man's Environmental Crisis* (Chicago: Rand McNally, 1968); Andrew Jamison, *The Steam-Powered Automobile: An Answer to Air Pollution* (Bloomington: Indian University Press, 1970); John C. Esposito, Ralph Nader, and Larry J. Silverman, *Vanishing Air: The Ralph Nader Study Group Report on Air Pollution* (New York: Grossman Publishers, 1970); E. J. Lehmann, *Automobile Air Pollution: A Bibliography with Abstracts* (Springfield: National Technical Information Service, 1973).

33. Ralph Nader, *L'auto che uccide*, trans. G. Boschi Breghelli (Milan: Bompiani, 1967). On the debate in Italy, see the section "Disciplina della circolazione stradale," in Pier Paolo Luzzatto Fegiz, *Il volto sconosciuto dell'Italia: Dieci anni di sondaggi Doxa* (Milan: Giuffrè, 1956), 590; Mino E. Damato, *L'auto che non uccide* (Rome: L'editrice dell'automobile, 1969); "È l'ora della sicurezza" and "Salvare la pelle," *Quattroruote*, May 1961.

34. Clyde L. King, *The Automobile: Its Province and Its Problems* (Philadelphia, 1924); Austin F. MacDonald, *Planning for City Traffic* (Philadelphia, 1927); Sir Herbert A. Tripp, *Town Planning and Road Traffic* (London: E. Arnold and Co., 1942); Robert B. Mitchell and Chester Rapkin, *Urban Traffic, a Function of Land Use* (New York: Columbia University Press, 1954); Colin Buchanan, *Traffic in Towns: A Study of the Long Term Problems of Traffic in Urban Areas* (London: HMSO, 1963); Kurt Leibbrand, *Transportation and Town Planning* (Cambridge, MA: MIT Press, 1964); Michael J. Thomson, *Great Cities and Their Traffic* (London: Gollancz, 1977). On Italy, see Paolo Ceccarelli, Bruno Gabrielli, and Renato Rozzi (eds.), *Traffico urbano: Che fare?* (Padua: Marsilio, 1968); Franco Gigli, *Urbanistica di settore: Il traffico* (Rome: Multistampa, 1965); Paolo Sica, *Traffico e forma urbana* (Florence: Coop. Libraria Usf, 1967); Pietro L. Fano and Marco Ponti, *Il traffico urbano in Italia* (Milan: FrancoAngeli, 1972); Nico Ventura, *Città e autostrade negli squilibri regionali* (Padua: Patron, 1978).

35. Sigfrido Ramírez Pérez, "Un'auto più pulita e sicura: La Comunità europea e l'avvio della standardizzazione per gli autoveicoli (1952–1970)," *I Frutti di Demetra* 21 (2009): 47–62.

36. Donatella Della Porta and Mario Diani, *Movimenti senza protesta? L'ambientalismo in Italia* (Bologna: il Mulino, 2004), 81–83. The authors also report data on the Antivivisection League (LAV; 13,500 members in 1997) and Greenpeace-Italia (40,000

members at the end of the 1990s). Thanks to the success of the referendum campaign, overall membership had increased to 475,180, 0.84% of the population.

37. D. Della Porta and M. Diani, *Movimenti senza protesta?* See also Nicola Zanardi, "Greenpeace: Obiettivo mass media," *Equilibri* 1, no. 1 (1997): 117–21; Enrico Chiarugi, "Legambiente: Slogan d'azione," *Equilibri* 1, no. 2 (1997): 287–91; Mariagrazia Midulla, "WWF: Prima i fatti poi le parole," *Equilibri* 1, no. 3 (1997): 431–37.

38. D. Della Porta and M. Diani, *Movimenti senza protesta?*, 79–80. See also Grant Jordan and William A. Maloney, *The Protest Business? Mobilizing Campaign Groups* (Manchester: Manchester University Press, 1997); Hein-Anton Van Der Heijden, "Political Opportunity Structure and the Institutionalization of the Environmental Movement," *Environmental Politics* 6, no. 4 (1997): 25–50.

39. Currently the main Italian environmental organizations are: Legambiente, WWF, Italia Nostra, Greenpeace, LIPU, Amici della Terra, Ente Nazionale Protezione Animali (ENPA), LAV, LAC, Federazione Nazionale Pro Natura, Verdi Ambiente e Società (VAS), and Marevivo. Legambiente—which calls itself "the most widespread Italian environmental association on the territory"—has 115,000 members and 1,000 local groups. The WWF (organized in 19 regional branches and in 200 territorial structures) does not supply any updated information about membership on its website: according to the 2007 year-end report there were 275,000 members. From the 2008 report, in which the association certified the proceeds deriving from memberships of €3,691,233.96, and assuming that all the associates pay a minimum of €20 (membership costs vary from the €20 of the Panda Explorers to €300 for Millennium Club members), we may presume an approximate number of around 185,000 members (there were 281,000 in 1999). Italia Nostra limits itself to giving information on its number of branches (205). Greenpeace (2009) declares 91,177 activists online and 41,714 donors. LIPU states that it has 42,000 supporters organized in 100 local branches. Amici della Terra makes public only its own organizational structure, subdivided into 8 regional associations and 61 local clubs. The FAI (organized in 14 regional secretariats and 112 delegations) declares 72,818 members (2008). LAV (organized in 85 local branches) has 40,000 supporters. ENPA (159 branches in 18 regions) has 45,487 members. LAC has 40 branches in 15 Italian regions (the European Federation against Hunting, to which LAC belongs, has 80,000 members in 14 European countries). The Federazione Nazionale Pro Natura has 117 associations. VAS is present in 15 regions with 68 circles. Marevivo is present in 10 regions with 29 local units.

40. De Lucia, *Se questa è una città*, 167–89.

41. De Lucia, *Se questa è una città*, 191.

42. Campos Venuti, *Città senza cultura*, 41–42.

43. Constitutional Law No. 3, Oct. 18, 2001. Previously, Presidential Decree No. 380, June 6, 2001, had entrusted to regional administrations "concurrent legislative power on the subject of building, in observance of the fundamental principles of state

legislation inferable from the provisions contained in the consolidated law," Art. 2. See Campos Venuti, *Città senza cultura*, 105–13.

44. Alessandra Goria, "Istituzioni per l'uso: UNEP," *Equilibri* 1, no. 1 (1997): 103–8.

45. In 1986 the Single Europe Act formally legitimized community environmental policies by introducing a new heading (VII) into the European Economic Community Treaty expressly dedicated to the environment. On European environmental policy, see Laura Scichilone, *L'Europa e la sfida ecologica: Storia della politica ambientale europea (1969–1998)* (Bologna: il Mulino, 2008).

46. Pellizzoni, *Conflitti ambientali*, 219.

47. Pellizzoni, *Conflitti ambientali*, 320.

CHAPTER 9: THE PARABOLA OF ITALIAN ENVIRONMENTALISM

1. The main exponents of original preservationism were Ralph Waldo Emerson (1803–1882), Henry David Thoreau (1817–1862), and John Muir (1838–1914). See especially Ralph W. Emerson, *Nature and Other Essays* (Mineola, NY: Dover Publications, 2009); Henry D. Thoreau, *Walden (Life in the Woods)/Civil Disobedience* (New York: Signet Classics, 2004); John Muir, *Nature Writings: The Story of My Boyhood and Youth, My First Summer in the Sierra, the Mountains of California, Stickeen; Selected Essays* (New York: Penguin Books, 1997); Michael P. Branch (ed.), *Reading the Roots: American Nature Writing before Walden* (Athens: University of Georgia Press, 2004); Lance Newman, *Our Common Dwelling: Henry Thoreau, Transcendentalism, and the Class Politics of Nature* (New York: Palgrave Macmillan, 2005); Terry Gifford, *Reconnecting with John Muir: Essays in Post-Pastoral Practice* (Athens: University of Georgia Press, 2006); Donald Worster, *A Passion for Nature: The Life of John Muir* (New York: Oxford University Press, 2008).

2. On Pinchot and conservationism, see Gifford Pinchot, *Breaking New Ground* (Washington, DC: Island Press, 1998); Char Miller, *Gifford Pinchot and the Making of Modern Environmentalism* (Washington, DC: Island Press, 2001); David Stradling (ed.), *Conservation in the Progressive Era: Classic Texts* (Seattle: University of Washington Press, 2004).

3. See Aldo Leopold, "The Conservation Ethic," in *Journal of Forestry* 31, no. 6 (1931): 634–43; Aldo Leopold, *A Sand County Almanac, and Sketches Here and There* (New York: Oxford University Press, 1949).

4. On environmental ideas, see Pascal Acot, *Storia dell'ecologia*, trans. Silvana Nesi Sirgiovanni (Rome: Lucarini, 1989); Donald Worster, *Storia delle idee ecologiche*, trans. Elisa Gunella (Bologna: il Mulino, 1994); Joseph E. De Steiguer, *The Origins of Modern Environmental Thought* (Tucson: University of Arizona Press, 2006); Mario Diani, *Isole nell'arcipelago: Il movimento ecologista in Italia* (Bologna: il Mulino, 1988), 50.

5. See Edgar H. Meyer, *I pionieri dell'ambiente* (Milan: Carabà, 1995), 19 and 72–78. Moreover, as Meyer reports, the 1870s saw the founding of the Société de la Flore

Valdôtaine and the Società Zoofila Piemontese (which in 1938 spawned ENPA, the national body for the protection of animals). Also active were the Società Botanica Italiana, the Gruppo Naturalistico "Giuseppe Ragazzoni," and the Club Alpino Italiano. On the birth of the protectionist movement in Italy, see also Fondazione Giovanni Agnelli, *Volontari per l'arte e per l'ambiente* (Turin: Fondazione Giovanni Agnelli, 1985); Luigi Piccioni, *Il volto amato della Patria: Il primo movimento per la protezione della natura in Italia 1880–1934* (Camerino: Università di Camerino, 1999); James Sievert, *The Origins of Nature Conservation in Italy* (New York: Peter Lang, 2000); Gianluigi Della Valentina, *Storia dell'ambientalismo in Italia: Lo sviluppo insostenibile* (Milan: Bruno Mondadori, 2011).

6. As Meyer, *I pionieri dell'ambiente*, 73–78, points out, the Touring Club promoted the establishment of a propaganda commission for woodland and pastures (1909) and subsequently a national committee for defense of the landscape and monuments (1913).

7. On these events, see Meyer, *I pionieri dell'ambiente*, 22–25, 30–39; Piccioni, *Il volto amato della Patria*, 126–38; Della Valentina, *Storia dell'ambientalismo in Italia*, 43–69.

8. See Giorgio Bassani, *Italia da salvare: Scritti civili e battaglie ambientali* (Turin: Einaudi, 2005), 5. See also Meyer, *I pionieri dell'ambiente*, 133–60; Antonio Cederna, *I vandali in casa* (Roma-Bari: Laterza, 1956).

9. On the origins of Pro Natura, see Meyer, *I pionieri dell'ambiente*, 41–46; and Luigi Piccioni, *Primo di cordata: Renzo Videsott dal sesto grado alla protezione della natura* (Trento: Tipografia Editrice Temi, 2010), 231–88.

10. On the Italian branch of the WWF, see Meyer, *I pionieri dell'ambiente*, 161–88.

11. Diani, *Isole nell'arcipelago*, 69.

12. Paul R. Ehrlich, *The Population Bomb* (New York: Ballantine Books, 1968); Garrett J. Hardin, "The Tragedy of the Commons," *Science* 162, no. 3859 (1968): 1243–48; Ezra J. Mishan, *The Costs of Economic Growth* (Harmondsworth, UK: Penguin, 1969); Herman E. Daly (ed.), *Essays toward a Steady-State Economy* (Cuernavaca: Centre Intercultural de Documentación, 1971); Edward Goldsmith and Robert Allen, *A Blueprint for Survival* (Harmondsworth, UK: Penguin, 1972).

13. In an influential magazine of Italian reformism, *il Mulino*, the first articles dedicated to environmental themes—apart from one dating to 1962, "Dialoghi di Cleante: Tutela del paesaggio e localizzazioni industriali," 119 (1962): 144–150—appeared in the two-year period 1972–1973: Carlo Cavallotti, "Costo, risanamento ecologico e sindacati dei lavoratori: Tempi e modi," 220 (1972): 374–82; Altiero Spinelli, "Una sfida per l'Europa: Lo sviluppo industriale e il problema ecologico," 221 (1972): 397–406; Massimo Roatti, "Etica ed estetica dell'ecologia: Aspettando il dies irae," 222 (1972): 710–36; Lynn T. White, "Le radici storico-culturali della nostra crisi ecologica," 226 (1973): 251–63; Leonardo Urbani, "Inquinamento di Stato?," 229 (1973): 849–62. Then only two articles in 1976 and 1983: Umberto Romagnoli, "Sui fatti di Seveso e di Padova,"

246 (1976): 643–47; and Angelo Bolaffi and Otto Kallscheuer, "I verdi: La teoria dei colori di un paradosso politico; Tra nuovi movimenti e mutamenti della politica," 288 (1983): 599–632. See *il Mulino, 1951–2000: Indice storico* (Bologna: il Mulino, 2001).

14. Vincenzo Caglioti, "L'uomo e l'ambiente," in Senato della Repubblica, *Problemi dell'ecologia*, vol. 1, *Nota introduttiva del presidente Fanfani ed illustrazioni dei professori Caglioti, Marini-Bettolo, Nebbia, Passino, Pavan e Tomaselli* (Rome: Tipografia del Senato, 1971), 2, 32.

15. The original edition was published in 1972 in New York by Universe Books, while in Italy it was published the same year by Edizioni scientifiche e tecniche Mondadori. On the Italian debate regarding the Club di Roma report, see Giorgio Nebbia and Luigi Piccioni, "Un tornante del dibattito italiano sull'ambiente: La ricezione dei Limiti dello sviluppo," *Ricerche Storiche* 41, no. 3 (2011): 519–40.

16. Tecneco, *Prima relazione sulla situazione ambientale del Paese*, 3 vols. (Rome: Tip. Carlo Colombo, 1973). On the importance of the *Prima relazione*, see Michele Caputo, "Un progetto per la salvaguardia del patrimonio ambientale italiano," *il Mulino* 56, no. 434 (2007): 1042–49.

17. Laura Conti, *Visto da Seveso: L'evento straordinario e l'ordinaria amministrazione* (Milan: Feltrinelli, 1977); A. Colombi, *Seveso due anni dopo: Materiale di documentazione* (Milan: Unicopli, 1978); Sergio Fusaro and Francesco P. Tronca, *Le attività industriali a rischio di incidente rilevante: Analisi e commento della direttiva Seveso* (Padua: Cedam, 1990); Bruno Ziglioli, *La mina vagante: Il disastro di Seveso e la solidarietà nazionale* (Milan: FrancoAngeli, 2010).

18. Rachel Carson, *Silent Spring* (Cambridge, MA: Riverside Press, 1962); Barry Commoner, *The Closing Circle: Nature, Man, and Technology* (New York: Knopf, 1971). See also these essential works: Serge Moscovici, *La Société contre nature* (Paris: Union générale d'éditions, 1972); Ivan Illich, *La convivialité* (Paris: Éditions du Seuil, 1973); Ernst F. Schumacher, *Small Is Beautiful: A Study of Economics as if People Mattered* (London: Blond and Briggs, 1973); Edgar Morin, *Le Paradigme perdu, la nature humaine* (Paris: Éditions du Seuil, 1973); André Gorz, *écologie et politique* (Paris: Galilée, 1975); René Dumont, *Seule une écologie socialiste* (Paris: R. Laffont, 1977).

19. Dario Paccino, *L'imbroglio ecologico: L'ideologia della natura* (Turin: Einaudi, 1972); Giovanni Ciccotti and Marcello Cini, *L'ape e l'architetto: Paradigmi scientifici e materialismo storico* (Milan: Feltrinelli, 1976); Laura Conti, *Che cos'è l'ecologia: Capitale, lavoro e ambiente* (Milan: G. Mazzotta, 1977); Virginio Bettini, *Autoimbroglio ecologico e limiti allo sviluppo* (Florence: Sansoni, 1979).

20. On the development of the Italian environmental movement, see Diani, *Isole nell'arcipelago*; Roberto Della Seta, *La difesa dell'ambiente in Italia: Storia e cultura del movimento ecologista* (Milan: FrancoAngeli, 2000); Della Valentina, *Storia dell'ambientalismo in Italia*. See also Roberto Della Seta and Daniele Guastini, *Dizionario del pensiero ecologico: Da Pitagora ai no-global* (Rome: Carocci, 2007).

21. The Amici della Terra was linked to Friends of the Earth, founded in 1969 in San Francisco by David Brower, then head of the Sierra Club. The new organization came into being in an open polemic with the Sierra Club which, on the occasion of the debate over the building of two nuclear power plants in California, decided not to take up a position against the use of atomic energy. On Friends of the Earth, see Centro Studi Investimenti Sociali (CENSIS), *Energia, sviluppo, ambiente* (Milan: Garzanti, 1990), 204–6; David R. Brower, *For Earth's Sake: The Life and Times of David Brower* (Salt Lake City, UT: Gibbs Smith Publisher, 1990).

22. "While the associations established before us have a prevalently conservationist and protectionist culture," explains Enrico Testa, the young national president of the association, "we came into being with criticism of the industrialist model of development, the crisis in natural and energy resources, the explosion of themes on the quality of life"; *Arcipelago verde: La prima guida completa per chi ama la natura*, Supplement to *Panorama*, no. 899 (July 5, 1983): 31.

23. See "Arcipelago verde," in *Arcipelago verde*, 58–59.

24. Pro Natura's (Federnatura) central structure was organized into five commissions dedicated to: protection of flora, parks and reserves; defense of the land and quarries and town planning; schools; hunting; fishing and water pollution. Mention should be made of the numerically small Kronos 1991 (500 members in the early 1980s), created at the end of the 1960s and active on specific campaigns such as forest fire prevention, promotion of solar energy, recycling of solid urban waste, and control of the quality of inland waterways (see http://www.accademiakronos.it, accessed Dec. 3, 2012).

25. The 76,000 members in 1983 were subdivided: 30,000 WWF, 17,000 LIPU, 15,000 Lega per l'ambiente, and 14,000 Italia Nostra. See *Arcipelago verde*, 23–65.

26. The referendum in the end was thrown out by the Constitutional Court; see Anna Chimenti, *Storia dei referendum: Dal divorzio alla riforma elettorale* (Milan: CDE, 1993), 49–69. On the incremental convergence of the political ecology and conservationist movements, see Mario Diani and Giovanni Lodi, "Three in One: Currents in the Milan Ecology Movement," *International Social Movements Research* 1 (1988): 103–24; Paolo R. Donati, "Building a Unified Movement: Resource Mobilization, Media Work, and Organizational Transformation in the Italian Environmentalist Movement," *Research in Social Movements, Conflicts and Change* 19 (1996): 125–57.

27. The referendum was also promoted by the Radical Party, Democrazia Proletaria, Sinistra Indipendente, FIGC (Italian Communist Youth Federation), and ARCI. On the referendum, see Chimenti, *Storia dei referendum*, 77–88; Saverio Luzzi, "Come l'Italia (non) uscì dal nucleare," *I Frutti di Demetra* 23 (2010): 45–65.

28. Donatella Della Porta and Mario Diani, *Movimenti senza protesta? L'ambientalismo in Italia* (Bologna: il Mulino, 2004), 81–83. The authors also report data relating

to the Antivivisection League (LAV; 13,500 members in 1997) and to Greenpeace-Italia (40,000 members at the end of the 1990s).

29. Diani, *Isole nell'arcipelago*, 58. Diani subdivides the organizations closest to the conservationist action model into groups of public interest (Italia Nostra, WWF, Federnatura, Kronos 1991, Amici della Terra), zoophile leagues (LIPU, LAC, LAV), civic and single-issue committees, and libertarian ecology groups (56–60).

30. Over and above Legambiente, Diani includes under political ecology the Green Lists (Federazione dei Verdi), the technical-scientific groups (Medicina Democratica, Geologia Democratica, AAM Terra Nuova), and the grassroots groups; see Diani, *Isole nell'arcipelago*, 60–61.

31. See CENSIS, *Energia, sviluppo, ambiente*, 206–9. "Mountain Wilderness activists are not afraid when required to have recourse also to spectacular and provocative actions, aware that only by regaining the ability to get indignant and not accept the fait accompli will it be possible to conserve and hand down to future generations the immense heritage that the mountains of the world have always offered us" (http://www.mountainwilderness.it, accessed Dec. 3, 2012).

32. The Gruppi di Ricerca Ecologica (GRE, Ecological Research Groups) were founded in 1978, close to the MSI political area that was more sensitive to the themes of communitarianism and environmentalism. Its first president was the biologist Alessandro Di Pietro, today a well-known TV presenter. Thus on the site of the Marche's Gruppi di Ricerca Ecologica: "The action of the Gruppi Ricerca Ecologica, based on volunteer work, on promotion of employment in the Environmental sector (Forestry, Management of nature oases and reserves, Food, Agriculture), on the safeguarding and positive exploitation of environmental and cultural heritage, and rooted in reality, in the needs of the Community, takes the form of a new policy of the environment and for the environment, which favors values. The values of Tradition, of Life, of the Person so mortified and annihilated today by criminal experimentation in the name of a poorly understood sense of progress. The GRE, which reaffirm the preeminence of spirit over matter, of being over having, of culture over the economy, of human dignity over the machine, intend to defend citizens from the mental pollution and deceitful directions—about customs and consumption—to which they are subjected through the distorting transformation of advertising, teaching and culture"; Diani, *Isole nell'arcipelago*, 61.

33. Diani, *Isole nell'arcipelago*, 66.

34. Della Porta and Diani, *Movimenti senza protesta?*, 79–80. See also Grant Jordan and William A. Maloney, *The Protest Business? Mobilizing Campaign Groups* (Manchester: Manchester University Press, 1997); Hein-Anton van der Heijden, "Political Opportunity Structure and the Institutionalization of the Environmental Movement," *Environmental Politics* 6, no. 4 (1997): 25–50.

35. Della Porta and Diani, *Movimenti senza protesta?*, 87–94. See also Christopher

Rootes (ed.), *Environmental Movements: Local, National and Global* (London: Frank Cass, 1999); Christopher Rootes, "The Transformation of Environmental Activism: Activists, Organizations and Policy-Making," *Innovation: The European Journal of Social Sciences* 12 (1999): 153–73; Mark Dowie, *Losing Ground: American Environmentalism at the End of the Twentieth Century* (Cambridge, MA: MIT Press, 1995).

36. See note 39 of chapter 8.

37. On environmental justice, see David Schlosberg, *Environmental Justice and the New Pluralism: The Challenge of Difference for Environmentalism* (New York: Oxford University Press, 1999); J. Donald Hughes (ed.), *The Face of the Earth: Environment and World History* (Armonk, NY: M. E. Sharpe, 2000); Richard Hofrichter (ed.), *Toxic Struggles: The Theory and Practice of Environmental Justice* (Salt Lake City: University of Utah Press, 2002); Marco Armiero, "Processi naturali: Conflitti ambientali e conflitti sociali tra XIX e XX secolo," in Giuseppe Civile and Giulio Machetti (eds.), *La città e il tribunale* (Naples: Dante e Descartes, 2004); Robert D. Bullard (ed.), *The Quest for Environmental Justice: Human Rights and the Politics of Pollution* (San Francisco: Sierra Club Books and University of California Press, 2005); David N. Pellow and Robert J. Brulle (eds.), *Power, Justice, and the Environment: A Critical Appraisal of the Environmental Justice Movement* (Cambridge, MA: MIT Press, 2006); Geneviève Massard-Guilbaud and Richard Rodger (eds.), *Environmental and Social Justice in the City: Historical Perspectives* (Cambridge: White Horse Press, 2011).

38. See William R. Freudenberg and Susan K. Pastor, "Nimbys and Lulus: Stalking the Syndromes," *Journal of Social Issues* 48, no. 3 (1992): 39–61; Carissa Schively, "Understanding the NIMBY and LULU Phenomena: Reassessing Our Knowledge Base and Informing Future Research," *Journal of Planning Literature* 21, no. 3 (2007): 255–66. On citizens' committees and the "environmental justice movement": Andrew Szasz, *Eco-populism: Toxic Waste and Movements for Environmental Justice* (London: UCL Press, 1994); Paul Lichtermann, *The Search for Political Community: American Activists Reinventing Tradition* (Cambridge: Cambridge University Press, 1996); Donatella Della Porta, "La democrazia partecipativa e i movimenti sociali: Micro e macro dinamiche," in Bianca R. Gelli (ed.), *Le nuove forme della partecipazione* (Rome: Carocci, 2007). On NIMBYism understood as an egoistic and particularistic stance, see Charles Piller, *The Fail-Safe Society: Community Defiance at the End of American Technological Optimism* (New York: Basic Books, 1991); Michael Dear, "Understanding and Overcoming the Nimby Syndrome," *Journal of the American Planning Association* 58, no. 3 (1992): 288–301; Eric Smith and Marisela Marquez, "The Other Side of the Nimby Syndrome," *Society and Natural Resources* 13, no. 3 (2000): 273–80; Phil Hubbard, "Accommodating Otherness: Anti-asylum Centre Protest and the Maintenance of White Privilege," *Transactions of the Institute of British Geographers* 30, no. 1 (2005): 52–65.

39. The case of the Federazione dei Verdi seems paradigmatic; having obtained

maximum consensus in 1992 (2.79% in the Chamber and 3.08% in the Senate), its appeal declined progressively in subsequent sessions (Chamber: 2.70% in 1994 and 2.50% in 1996) until it diluted into the modest results shown by the election cartels in the 2000s (Chamber: 2.17% of Il Girasole, together with the SDI [Italian Democratic Socialists], in 2001; and 3.08% of La Sinistra/l'Arcobaleno, with Rifondazione Comunista, Partito dei Comunisti Italiani, Sinistra Democratica, in 2008).

40. According to the website of Rete dei Comitati ("Who We Are," http://www.territorialmente.it, accessed Feb. 21, 2013) in 2007, there were 114 member committees, increasing to 166 in 2009. The active supporters—the site does not specify the year—were 1,682, with 16,818 participating supporters (0.46% of the Tuscan population on January 1, 2007). Florence Province clearly prevailed in geographical distribution (44.74% in 2007 and 40.36% in 2009). Last, personal data demonstrate that Rete dei Comitati attracted mainly citizens aged on average over 45: in fact in five provinces the average age was 50 or over (Florence, Livorno, Pisa, Pistoia, Prato), in three it was 45 or over (Grosseto, Siena, Lucca), in one between 40 and 50 (Arezzo), and under 40 only in Massa.

41. Emilia Blanchetti, Silvia Capotorto, and Emilio Conti, "Fenomenologia della sindrome NIMBY," online supplement *Limes: Rivista italiana di geopolitica*, no. 6 (2007), http://limes.espresso.repubblica.it/2007/11/22/il-clima-dellenergia-20/?p=359; Giorgio Osti, "Come inquadrare i conflitti ambientali," in Duccio Bianchi and Giuseppe Gamba (eds.), *Ambiente Italia 2007: Rapporto annuale di Legambiente* (Milan: Edizioni Ambiente, 2007), 15–31.

42. Osservatorio NIMBY Forum, 2005–2006, www.nimbyforum.it. Waste management: waste-to-energy plant (32.37% involving 56 plants), incinerators (1.15%, 2), composting-treatment plants (9.25%, 16), garbage dumps (10.41%, 18), nuclear waste storage sites (0.58%, 1). Production and distribution of energy from nonrenewable sources: thermoelectric power plants (14.45%, 25), power lines (4.63%, 8), LNG terminals (3.47%, 6), gasifiers (1.73%, 3), gas pipelines (1.73%, 3). Transportation infrastructure: roads (6.94%, 12), rail lines (2.90%, 5), airports (1.15%, 2), other (1.15%, 2). Production of energy from renewable sources: biomass power plants (5.20%, 9), wind farms (1.73%, 3), hydroelectric power plants (0.58%, 1). Other infrastructures (0.58%, 1). Data were drawn from Blanchetti, Capotorto, and Conti, "Fenomenologia della sindrome NIMBY."

43. Between 2005–2006 and 2010–2011 the episodes of opposition to the construction of plants and infrastructure rose from 102 to 163 in the north (+59.80%), from 34 to 92 in the center (+170.59%), and from 37 to 89 in the south (+140.54%). Regional distribution 2005–2006: North: Emilia Romagna, 15; Friuli Venezia Giulia, 9; Liguria, 3; Lombardy, 43; Piedmont, 12; Trentino Alto Adige, 4; Veneto, 16. Center: Lazio, 10; Marche, 4; Tuscany, 17; Umbria, 3. South: Abruzzo, 5; Basilicata, 1; Calabria, 2; Campania, 6; Molise, 1; Puglia, 11; Sardinia, 3; Sicily, 8. Regional distribution 2010–

2011: North: Emilia Romagna, 31 (+106.66%); Friuli, 18 (+100%); Liguria, 8 (+166.66%); Lombardy, 46 (+6.97%); Piedmont, 8 (–33.33%); Trentino Alto Adige, 11 (+175%); Val d'Aosta, 1 (+100%); Veneto, 40 (+150%). Center: Lazio, 19 (+90%); Marche, 20 (+400%); Tuscany, 42 (+147.06%); Umbria, 11 (+266.66%). South: Abruzzo, 22 (+340%); Basilicata, 5 (+500%); Calabria, 11 (+450%); Campania, 15 (+150%); Molise, 2 (+100%); Puglia, 10 (–9.09%); Sardinia, 11 (+266.66%); Sicily, 12 (+50%). Data were drawn from http://www.nimbyforum.it, accessed Dec. 4, 2012.

44. Production of energy from renewable sources: biomass power plants (24.42% involving 84 plants), wind farms (12.50%, 43), hydroelectric power plants (6.98%, 24), photovoltaic systems (1.74%, 6). Waste management: garbage dumps (13.09%, 45), waste-to-energy plant (9.59%, 33), composting-treatment plants (4.07%, 14), incinerators (0.29%, 1), hazardous waste treatment plants (0.29%, 1). Production and distribution of energy from nonrenewable sources: thermoelectric power plants (5.82%, 20), power lines (4.94%, 17); gas pipelines (3.77%; 13), gasifiers (2.33%, 8), LNG terminals (2.33%, 8), natural gas pipelines (0.87%, 3), gas extraction plants (0.29%, 1), gas storage plants (0.29%, 1). Transportation infrastructure: roads (3.20%, 11); rail lines (0.58%, 2); airports (0.29%, 1), other infrastructure (2.32%, 8). Data were drawn from http://www.nimbyforum.it, accessed Dec. 4, 2012.

45. Pietro Menna, *L'energia pulita: Sole, acqua, vento . . . le fonti di energia rinnovabili* (Bologna: il Mulino, 2003), 85. See also G8 Renewable Energy Task Force, *Chairmen's Report*, G8 Information Centre, 2001, www.g8.utoronto.ca; Lester R. Brown, *Piano B 3.0: Mobilitarsi per salvare la civiltà*, trans. Attilio Anitori (Milan: Edizioni Ambiente, 2008), 251–73.

46. For example, an explicit break occurred over the construction of an incinerator on the Florence plain, where the WWF opposed it categorically while Legambiente accepted it as long as there were efficient recycling policies. See Silvia Givone, "L'inceneritore della piana fiorentina: Una decisione senza partecipazione," in Massimo Morisi and Andrea Paci (eds.), *Il bisogno di decidere* (Bologna: il Mulino, 2009), 189–205. A further example concerns the creation of the Poggi Alti wind farm near Scansano in Tuscany (ten wind turbines for an overall yield of 10 MW). Opposition was headed by the winegrowers of the area (where Morellino is produced) and by Italia Nostra, while Legambiente and the WWF expressed a substantially favorable opinion. See "All'assalto della Toscana i don Chisciotte dell'Anev," *La Repubblica*, May 26, 2008; "Viticoltori contro il parco eolico: Cambia clima e paesaggio," *La Repubblica*, May 26, 2005; "La battaglia dei mulini a vento," *La Repubblica*, May 12, 2008.

47. Anti-utilitarianism—born in France in the context of anthropological criticism of the economy and inspired by the works of David Émile Durkheim, Marcel Mauss, Karl Paul Polanyi, and Marshall Sahlins—is a current of contemporary sociological thought which upholds the urgency of going beyond the economics characterizing today's societies in order to give rise to a profound social renewal centered on the fair

distribution of resources and on solidarity. See Serge Latouche, *L'occidentalizzazione del mondo: Saggio sul significato, la portata e i limiti dell'uniformazione planetaria*, trans. Alfredo Salsano (Turin: Bollati Boringhieri, 1992); Serge Latouche, *Breve trattato sulla decrescita serena*, trans. Fabrizio Grillenzoni (Turin: Bollati Boringhieri, 2007); Maurizio Pallante, *La decrescita felice: La qualità della vita non dipende dal Pil* (Rome: Editori Riuniti, 2005).

48. The term *organicism* indicates all those doctrines which tend to interpret the world and nature by analogy with the organism. The organicist theory most cited by radical organizations is the "Gaia hypothesis," elaborated by the chemist James Lovelock, which considers the Earth as a single living organism capable of self-regulation. The Gaia hypothesis has three basic principles: the first maintains that there is a bidirectional relationship between the physical environment and living organisms as a result of which evolution comes to be the product of interaction between the matrices of the biosphere (atmosphere, soils, oceans) and biotic communities; the second overturns Darwinian assumption by stating that the survival of organisms is ensured not by mechanisms of competition but by natural forms of cooperation; the third asserts that evolution of the biosphere is a process capable of influencing the earth's climate and the physical-chemical conditions necessary for maintaining life. See James Lovelock, *Gaia: Nuove idee sull'ecologia*, trans. Vania Bassan Landucci (Turin: Boringhieri, 1981); James Lovelock, *La rivolta di Gaia*, trans. Massimo Scaglione (Milan: Rizzoli, 2006).

49. Deep ecology is a current of thought begun by the Norwegian philosopher Arne Naess which maintains that it is necessary to go beyond anthropocentrism inasmuch as all living beings have an identical intrinsic importance (i.e., independent of calculations of an economic or practical order) and cannot be classified on the basis of their relative value. See Arne Naess, *Ecology, Community, and Lifestyle: Outline of an Ecosophy* (Cambridge: Cambridge University Press, 1989); Alan Drengson and Yuichi Inoue (eds.), *The Deep Ecology Movement: An Introductory Anthology* (Berkeley: North Atlantic Books, 1995); Guido Dalla Casa, *Ecologia profonda* (Turin: Pangea, 1996).

50. Bioregionalism is a practice of ecological life centered on the concept of the bioregion, understood as a geographical place recognizable by its characteristics of soil, vegetable and animal species and climate, as well as for the human culture which since time immemorial has developed in harmony with all of this. Bioregionalism is therefore characterized as a movement of autonomism which aims to overcome the ecological crisis by advocating the need for a new socioeconomic system of community and regional types, ruled by local government. See Daniel Berthold-Bond, "The Ethics of Place: Reflections on Bioregionalism," *Environmental Ethics* 22, no. 1 (2000): 5–24; Emanuele Guerrieri Ciacieri, *Bioregionalismo: La visione locale di un mondo globale* (Ragusa: Argo, 2006).

51. "Let us imagine the number of jobs created if we move from a depredation

economy to an ecosustainable system. Lester Brown . . . identifies nine sectors of production which should be developed within a 'solar' economy, which is to say, founded on renewable energy: the construction of wind farms and related turbines, the production of photovoltaic panels, the bicycle industry, the production of hydrogen and the related engines, the construction of lightweight underground railways, organic agriculture and reforestation activities"; Serge Latouche, *La scommessa della decrescita*, trans. Matteo Schianchi (Milan: Feltrinelli, 2009), 126.

52. Agenda 21 was a UN program established in 1992 during the Conference on Environment and Development in Rio de Janeiro, whose literal meaning is a list of appointments for achieving sustainable development in the course of the twenty-first century. In Europe it was actively promoted by the Aalborg Charter (May 27, 1994)—through which eighty municipalities prepared long-term plans for lasting and sustainable development—but not even it appeared capable of actively involving citizens in decision-making processes concerning environmental questions.

53. In cases like this some scholars speak of NIABY (Not in Any Backyard) or of NOTPE (Not on the Planet Earth). See Gregory E. Mcavoy, "Partisan Probing and Democratic Decision Making: Rethinking the Nimby Syndrome," *Policy Studies Journal* 26, no. 2 (1998): 274–92; and Maarten Wolsink, "Wind Power and the Nimby-Myth: Institutional Capacity and the Limited Significance of Public Support," *Renewable Energy* 21, no. 1 (2000): 49–64.

54. Luigi Bobbio, "Democrazia e nuove forme di partecipazione," in Michelangelo Bovero and Valentina Pazé (eds.), *La democrazia in nove lezioni* (Roma-Bari: Laterza, 2010), 46–63. See also Giovanni Allegretti, *L'insegnamento di Porto Alegre: Autoprogettualità come paradigma urbano* (Florence: Alinea, 2003); James S. Fishkin, *La nostra voce: Opinione pubblica e democrazia, una proposta* (Venice: Marsilio, 2003).

55. "One of the most glaring manifestations of the crisis of democracy may be recognized in the spreading on a planetary scale of certain forms of political action that some scholars have baptized with the neologism 'antipolitics.' Although the concept is still nebulous, the term designates with fair approximation the vision and strategies of parties and movements that aim to gather consensus around demagogic neo-populist formulas . . . Well, to designate both, those of the right and of the pseudo-left, I would be tempted to adopt, instead of the neologism 'antipolitics,' the more explicit term 'antidemocracy'; also to suggest that in spite of the electoral consensus obtained by these political players, we are dealing with a caricature, indeed an aping of democracy: with an apparent democracy dressed and disguised in incipient forms of elective autocracy"; Michelangelo Bovero, "Democrazia al crepuscolo?," in Bovero and Pazé, *La democrazia*, 17. On antipolitics, see Donatella Campus, *L'antipolitica al governo: De Gaulle, Reagan, Berlusconi* (Bologna: il Mulino, 2006); Fabio Bordignon, *L'Europa unita . . . dall'antipolitica: Società, politica e partiti nell'Europa post-comunista*

(Naples: Liguori, 2009); Leonardo Facco, *Elogio dell'antipolitica: In difesa delle libertà individuali* (Soveria Mannelli: Rubbettino, 2012).

56. Tony Judt, *L'età dell'oblio: Sulle rimozioni del '900*, trans. Paolo Falcone (Roma-Bari: Laterza, 2011), 409.

57. Judt, *L'età dell'oblio*, 409.

58. Civic committees do not escape the paradox of participatory practices: "in theory they address all citizens without exception, but in practice they manage to involve only a small (often very small) minority"; Luigi Bobbio, "Democrazia e nuove forme di partecipazione," 58.

CHAPTER 10: POPULAR INITIATIVE MOVEMENTS AND THE COMMONS

1. Francesco Tuccari, "Democrazie acefale e dispotismo postdemocratico," *Storia del pensiero politico* 1, no. 1 (2012): 105–41. The literature on the cause of the crisis in representative democracy is now quite vast; see primarily Robert A. Dahl, *Polyarchy: Participation and Opposition* (New Haven, CT: Yale University Press, 1971); Francis Fukuyama, *The End of History and the Last Man* (New York: The Free Press, 1992); Danilo Zolo, *Il principato democratico* (Milan: Feltrinelli, 1992); Thomas Piketty, *L'économie des inégalités* (Paris: La Découverte & Syros, 1997); Giovanni Sartori, *Homo videns: Televisione e post-pensiero* (Roma-Bari: Laterza, 1997); Zygmunt Bauman, *Globalization: The Human Consequences* (Cambridge: Polity Press; Oxford: Blackwell, 1998); Ulrich Beck, *Was ist Globalisierung? Irrtümer des Globalismus-Antworten auf Globalisierung* (Frankfurt am Main: Suhrkamp Verlag, 1997); Robert A. Dahl, *On Democracy* (New Haven: Yale University Press, 1998); Ulrich Beck, *Macht und Gegenmacht im globalen Zeitalter: Neue weltpolitische Ökonomie* (Frankfurt am Main: Suhrkamp Verlag, 2002); Massimo L. Salvadori, *Democrazie senza democrazia* (Roma-Bari: Laterza, 2009); Tony Judt, *Ill Fares the Land* (New York: Penguin Press, 2010); Richard Heinberg, *The End of Growth: Adapting to Our New Economy Reality* (Gabriola Island, BC: New Society Publishers, 2011); Carlo Altini (ed.), *Democrazia: Storia e teoria di un'esperienza filosofica e politica* (Bologna: il Mulino, 2011); Raghuram G. Rajan, *Fault Lines: How Hidden Fractures Still Threaten the World Economy* (Princeton, NJ: Princeton University Press, 2010); Thomas Piketty, *Le capital au XXI siècle* (Paris: Editions du Seuil, 2013).

2. Tuccari, "Democrazie acefale e dispotismo postdemocratico," 136.

3. Christian Salmon, *La politica nell'era dello storytelling*, trans. Nicola Vincenzoni (Rome: Fazi Editore, 2014), 9. See also Mauro Calise, *Il partito personale: I due corpi del leader* (Roma-Bari: Laterza, 2000); Ilvo Diamanti, "La democrazia degli interstizi: Società e partiti in Europa dopo la caduta del Muro," *Rassegna italiana di sociologia*, no. 3 (2007): 387–411.

4. Tuccari, "Democrazie acefale e dispotismo postdemocratico," 106–7, 141. On the

poor governing capacity of present-day ruling classes, see Slavoj Žižek, *Welcome to Interesting Times!* (London: Verso Books, 2010).

5. Donatella Della Porta and Mario Diani, *Movimenti senza protesta? L'ambientalismo in Italia* (Bologna: il Mulino, 2004), 79–80. See also Grant Jordan and William A. Maloney, *The Protest Business? Mobilizing Campaign Groups* (Manchester: Manchester University Press, 1997); Hein-Anton van der Heijden, "Political Opportunity Structure and the Institutionalization of the Environmental Movement," *Environmental Politics* 6, no. 4 (1997): 25–50.

6. Della Porta and Diani, *Movimenti*, 87–94. See also Christopher Rootes (ed.), *Environmental Movements: Local, National and Global* (London: Frank Cass, 1999); Christopher Rootes, "The Transformation of Environmental Activism: Activists, Organizations and Policy-Making," *Innovation: The European Journal of Social Sciences* 12 (1999): 153–73; Mark Dowie, *Losing Ground: American Environmentalism at the End of the Twentieth Century* (Cambridge, MA: MIT Press, 1995).

7. For a summary of the debate on this concept, see Ugo Mattei, *Beni comuni: Un manifesto* (Roma-Bari: Laterza, 2011); and Ermanno Vitale, *Contro i beni comuni: Una critica illuminista* (Roma-Bari: Laterza, 2013).

8. Paolo Cacciari (ed.), *La società dei beni comuni: Una rassegna* (Rome: Ediesse, 2010), 12–13. See also Gregorio Arena and Christian Iaione, *L'Italia dei beni comuni* (Rome: Carocci, 2012).

9. Cacciari, *La società*, 12–13.

10. The economic-environmental approach is adopted in international policies aimed at rendering development sustainable. The economy of the environment derives from the economy of well-being founded by Pigou and Mishan and deals with analyzing the economic aspects of environmental policies and their effects on collectivity. See Arthur C. Pigou, *The Economics of Welfare* (London: Macmillan, 1920); Ezra J. Mishan, *The Costs of Economic Growth* (New York: Praeger, 1967); Alessandro Vercelli and Simone Borghesi, *La sostenibilità dello sviluppo globale* (Rome: Carocci, 2005); Nicholas Stern, *A Blueprint for a Safer Planet: How to Manage Climate Change and Create a New Era of Progress and Prosperity* (London: Bodley Head, 2009); Jean-Paul Fitoussi and Eloi Laurent, *La nouvelle écologie politique: Economie et développement humain* (Paris: éditions du Seuil, 2008). As for the monetization of natural resources, this is considered a tool of fundamental importance to contain global environmental problems such as climate change. See Carlo Carraro and Alessandra Mazzai, *Il clima che cambia: Non solo un problema ambientale* (Bologna: il Mulino, 2015), 183–86.

11. Officina delle idee di Rete@Sinistra, "La società dei beni comuni," in Cacciari, *La società*, 17–26.

12. Mattei, *Beni comuni*, 100–101.

13. Mattei, *Beni comuni*, 104–5.

14. Mario Pirani, "All'assalto della Toscana i don Chisciotte dell'Anev," *La Repubblica*, May 26, 2008.

15. Pirani (who was a functionary of Mattei's ENI) has shown himself to be one of the main critics of wind energy but has supported shale gas, a far more environmentally impactful source. See the following articles that appeared in *La Repubblica*: "Il vento soffia miliardi a scapito del paesaggio," Mar. 17, 2008; "E se barbarie genera nel vento nuovi mostri," May 19, 2008; "La rivoluzione del gas quasi da nessuno percepita," Dec. 20, 2010; "I professori sul nucleare non sempre l'imbroccano," Apr. 4, 2011.

16. On the referendum, see "Firenze: Vincono i no alla tramvia; Referendum consultivo sulle linee 2 e 3; Contrari poco più del 50% dei votanti," *Corriere della Sera*, Feb. 18, 2008.

17. I produttori del Morellino contro il parco eolico di Scansano insieme all'associazione ambientalista Italia Nostra, Wine Report, Apr. 28, 2006, http://www.winereport.com: "Everyone agrees that it would be disastrous to implement a project which will have a powerful impact not only on the environment but also on the lives of the population and on tourism, with fallout on the image of the entire zone which for some years has been reevaluated from a tourism viewpoint, thanks also to wine producers who have invested millions of euros here, planting vineyards and building cellars that are subject to strict landscape restrictions. According to extensive research carried out in Germany, in areas close to wind farms the value of land falls drastically, while the flow of tourists slows down or is wholly interrupted."

18. In Florence the quality of the air remains one of its main environmental problems. See "Smog, lunedì nero in Toscana nelle centrali pm_{10} alle stelle," *La Repubblica*, Dec. 2, 2015; "Primo stop al traffico più vicino se lo smog non cala," *La Repubblica*, Dec. 22, 2015; "La direttrice Arpat ai sindaci toscani: Uno smog così non si era mai visto, qualcosa ora dovrete pur fare," *La Repubblica*, Dec. 27, 2015; "Toscana assediata dallo smog. L'Arpat: ora misure eccezionali," *Corriere della Sera*, Dec. 28, 2015.

19. James R. Flynn, *Senza alibi: Il cambiamento climatico; Impedire la catastrofe*, trans. Libero Sosio (Turin: Bollati Boringhieri, 2015), 142.

20. Stefania Ferrando, "Una politica dell'avvenire: Conversazione con Bruno Latour," *Equilibri*, no. 2 (2015): 335–45.

21. On the Rete, consult the website www.territorialmente.it.

22. Dario Giardi and Valeria Trapanese, *Dizionario dell'ambiente* (Florence: Alinea Editrice, 2007), 30–31.

23. On the complex meanings of the term *nature*, see Roberto Bondì and Antonello La Vergata, *Natura* (Bologna: il Mulino, 2015). See also Massimo Cacciari, "La pretesa faustiana dell'ecologismo," *Equilibri*, no. 3 (1998): 337–41; Philippe Descola, "Le nature del mondo," *Equilibri*, no. 3 (2006): 469–87; Chicco Testa and Patrizia Feletig, *Contro (la) natura: Perché la natura non è buona, né giusta, né bella* (Venice: Marsilio, 2014).

24. DOP stands for Denominazione d'Origine Protetta (Protected Designation of

Origin). This mark is guaranteed by the European Union and was created to promote the authenticity and artisan characteristics of certain food and agricultural products. These products are split into the categories of cheese, fruit and vegetables, cured meats, and olive oils. A DOP certification guarantees that a product and all phases of production for that product have been carried out in a strictly defined geographic area. With some products, this could be just a small cluster of villages. Although some are easily recognized, such as *prosciutto di Parma, aceto balsamico tradizionale di Modena,* and *Parmigiano Reggiano,* others are not so commonly known, such as *castagna del Monte Amiata* (a chestnut from Monte Amiata, Tuscany) or Castelmagno cheese from Piedmont.

25. Highly interesting on this subject is Valerio Neri's essay, "Fai, modelli organizzativi e dialettica istituzionale di un'associazione post-ambientalista," *Equilibri,* no. 2 (2004): 275–276: "The FAI forcefully reaffirms a well-known opinion: that human action always and necessarily implies a transformation of the natural environment, and therefore a possible, hoped-for integration. Art, architecture . . ., crafts, etc. are the creators of such happy integrations. . . . But at this point FAI's relationship to environmentalism seems to get lost. But no, the relationship is maintained; but it should be understood. . . . The FAI in fact believes that in the past, until the 20th century, this integration had at least two essential features: a) effective sustainability with regard to the natural environment, i.e. an acceptable impact on the fundamental mechanisms of nature, not devastating; b) the conviction that each historical period was born from the previous one in a line of development that continually innovated and integrated, binding itself intimately to the reality of the local population. Well, none of the two features is proper to the contemporary model of economic development, which is globalized and massifying, and it is that very model which, according to environmentalism, is guilty of devastating Nature!."

26. With regard to interpretations favorable to the LULU movements, and for a reference bibliography, see Michele Roccato and Terri Mannarini, *Non nel mio giardino: Prendere sul serio i movimenti NIMBY* (Bologna: il Mulino, 2012). See also David N. Pellow, "Framing Emerging Environmental Movement Tactics: Mobilizing Consensus, Demobilizing Conflict," *Sociological Forum* 14, no. 4 (1999): 659–83; Maria Kousis, "Economic Opportunities and Threats in Contentious Environmental Politics: A View from the European South," *Theory and Society* 33, no. 3–4 (2004): 393–415; Alessandra Algostino, "La democrazia e le sue forme: Una riflessione sul movimento No Tav," *Politica del diritto,* no. 4 (2007): 653–702; Lorena Rocca, "Partecipazione come pratica territoriale," *Equilibri,* no. 1 (2007): 117–24.

27. See note 40 of chapter 9.

28. "One of the most glaring manifestations of the crisis of democracy may be recognized in the spreading on a planetary scale of certain forms of political action that some scholars have baptized with the neologism 'antipolitics.' Although the

concept is still nebulous, the term designates with fair approximation the vision and strategies of parties and movements that aim to gather consensus around demagogic neo-populist formulas. . . . Well, to designate both, those of the right and of the pseudo-left, I would be tempted to adopt, instead of the neologism 'antipolitics,' the more explicit term 'antidemocracy'; also, to suggest that in spite of the electoral consensus obtained by these political players, we are dealing with a caricature, indeed an aping of democracy, with an apparent democracy dressed and disguised in incipient forms of elective autocracy"; Michelangelo Bovero, "Democrazia al crepuscolo?," in Michelangelo Bovero and Valentina Pazé (eds.), *La democrazia in nove lezioni* (Roma-Bari: Laterza, 2010), 17.

29. Tony Judt, *L'età dell'oblio: Sulle rimozioni del '900* (Roma-Bari: Laterza, 2011), 409.

30. Judt, *L'età dell'oblio.*

31. Elisa Olivito, "Partecipazione e ambiente: Una sguardo critico," *Diritto Pubblico*, no. 2 (2011): 559–610, 604, 605, 608. See also Luigi Bobbio, "Il dibattito pubblico sulle grandi opere: Il caso dell'autostrada di Genova," *Rivista italiana di politiche pubbliche*, no. 1 (2010): 119–46.

BIBLIOGRAPHY

ACI (Automobile Club of Italy). *XXX Conferenza del traffico e della circolazione: Relazioni e comunicazioni; Stresa, 27–30 settembre 1973.* Rome, 1973.

Acot, Pascal. *Storia dell'ecologia.* Translated by Silvana Nesi Sirgiovanni. Rome: Lucarini, 1989.

Adorno, Salvatore, and Simone Neri Serneri, eds. *Industria, ambiente e territorio: Per una storia ambientale delle aree industriali in Italia.* Bologna: il Mulino, 2009.

Adorno, Salvatore, and Simone Neri Serneri. "Per una storia ambientale delle aree industriali in Italia: Introduzione." In *Industria, ambiente e territorio,* edited by Salvatore Adorno and Simone Neri Serneri, 13–29. Bologna: il Mulino, 2009.

Agenzia per la Protezione dell'Ambiente. *Qualità dell'ambiente urbano.* Rome: APAT, 2006.

Algostino, Alessandra. "La democrazia e le sue forme: Una riflessione sul movimento No Tav." *Politica del diritto,* no. 4 (2007): 653–702.

Allan, John A. "'Virtual Water': An Essential Element in Stabilizing the Political Economies of the Middle East." *Yale University Forestry & Environmental Studies Bulletin,* no. 103 (1998): 141–49.

Allegretti, Giovanni. *L'insegnamento di Porto Alegre: Autoprogettualità come paradigma urbano.* Florence: Alinea, 2003.

Allen, Barbara L. *Uneasy Alchemy: Citizens and Experts in Louisiana's Chemical Corridor Disputes.* Cambridge, MA: MIT Press, 2003.

Altini, Carlo, ed. *Democrazia: Storia e teoria di un'esperienza filosofica e politica.* Bologna: il Mulino, 2011.

Amendola, Gianfranco. *Inquinati e inquinatori.* Rome: La Nuova Italia Scientifica, 1980.

Amendola, Gianfranco, and Paolo Dell'Anno. *La disciplina dell'inquinamento atmosferico.* Naples: Jovene Editore, 1992.

American Association for the Advancement of Science. Committee on Science in the Promotion of Human Welfare. Air Conservation Commission. *Air Conservation Report.* Washington, DC, 1965.

ANFIA (Associazione Nazionale fra Industrie Automobilistiche e Affini). *L'automobile in cifre.* Torino, 1950, 1951, 1952.

Appadurai, Arjun, ed. *The Social Life of Things: Commodities in Cultural Perspectives.* Cambridge: Cambridge University Press, 1986.

Arcipelago verde: La prima guida completa per chi ama la natura. Supplement to *Panorama,* no. 899 (July 5, 1983).

Arena, Gregorio, and Christian Iaione. *L'Italia dei beni comuni.* Rome: Carocci, 2012.

Armiero, Marco. "Processi naturali: Conflitti ambientali e conflitti sociali tra XIX e XX secolo." In *La città e il tribunale*, edited by Giuseppe Civile and Giulio Machetti. Naples: Dante e Descartes, 2004.

ARPAT (Agenzia Regionale per la Protezione Ambientale della Toscana). *Segnali ambientali in Toscana 2001.* Florence: Edifir, 2001.

ARPAT. *La qualità delle acque marine costiere in Toscana.* Florence: Edifir, 2004.

ARPAT. *Segnali ambientali in Toscana 2005.* Florence: Edifir, 2005.

ARPAT. *Relazione sullo stato dell'ambiente in Toscana 2008.* Florence: Edifir, 2008.

ARPAT. *Segnali ambientali in Toscana 2009.* Florence: Edifir, 2009.

Arrigoni, Tiziano, and Marisa Giachi, eds. *Identità di una città industriale: Piombino 1860–1992.* Piombino: Comune di Piombino, 1998.

Autorità di Bacino del Fiume Arno. *Trasformazioni del territorio e sviluppo dell'edificato lungo il corso dell'Arno e degli affluenti (1954–1993 e 1995).* Florence, 1997.

Aylesworth, Thomas G. *This Vital Air, This Vital Water: Man's Environmental Crisis.* Chicago: Rand McNally, 1968.

Bai, Xuemei. "Rizhao: Una città a energia solare." In Worldwatch Institute, *State of the World 2007: Il nostro futuro urbanizzato*, 242–44. Milan: Edizioni Ambiente, 2007.

Barberini, Mario. *Scarlino e il suo territorio nella evoluzione storica della Maremma.* Pisa: Nistri-Lischi, 1985.

Barducci, Gianluca, and Giorgio Casule. *Fanghi rossi di Scarlino, dieci anni di esperienze: L'evoluzione produttiva e ambientale dell'unico stabilimento italiano di produzione di biossido di titanio.* Florence: Regione Toscana, 1982.

Barles, Sabine. *L'invention des déchets urbains: France 1790–1970.* Seyssel: Champ Vallon, 2005.

Barnard, Alex V. *Freegans: Diving into the Wealth of Food Waste in America.* Minneapolis: University of Minnesota Press, 2016.

Barthes, Roland. *Scritti: Società, testo, comunicazione.* Translated by Gianfranco Marrone. Milan: CDE, 1998.

Bassani, Giorgio. *Italia da salvare: Scritti civili e battaglie ambientali.* Turin: Einaudi, 2005.

Bassoli, Romeo, Calogero Messana, and Roberto Vigotti. *Energia dal sole: Prospettive dell'energia fotovoltaica in Italia.* Rome: Hypothesis, 1992.

Battilani, Patrizia. *Vacanze di pochi, vacanze di tutti: L'evoluzione del turismo europeo.* Bologna: il Mulino, 2001.

Baudrillard, Jean. *La société de consomation: Ses mythes, ses structures.* Paris: Gallimard, 1974.

Bauman, Zygmunt. *Globalization: The Human Consequences.* Cambridge: Polity Press; Oxford: Blackwell, 1998.

Becattini, Giacomo. "Riflessioni sullo sviluppo socio-economico della Toscana in questo dopoguerra." In *Storia d'Italia: Le Regioni dall'unità a oggi, La Toscana*, edited by Georgio Mori. Turin: Einaudi, 1986.

Beck, Ulrich. *Was ist Globalisierung? Irrtümer des Globalismus-Antworten auf Globalisierung.* Frankfurt am Main: Suhrkamp Verlag, 1997.

Beck, Ulrich. *Macht und Gegenmacht im globalen Zeitalter: Neue weltpolitische* Ökonomie, Frankfurt am Main: Suhrkamp Verlag, 2002.

Benidickson, Jamie. *The Culture of Flushing: A Social and Legal History of Sewage.* Vancouver: UBC Press, 2007.

Berdini, Paolo. *Breve storia dell'abuso edilizio in Italia: Dal ventennio fascista al prossimo futuro.* Rome: Donzelli, 2010.

Berman, Daniel M., and John T. O'Connor. *Who Owns the Sun? People, Politics and the Struggle for a Solar Economy.* White River Junction, VT: Chelsea Green Publishing, 1996.

Bernhardt, Christopher, ed. *Environmental Problems in European Cities in the 19th and 20th Centuries.* Munich: Waxman, 2001.

Bernhardt, Christopher, and Geneviève Massard-Guilbaud, eds. *Le démon moderne—The modern Demon: La pollution dans les sociétés urbaines et industrielles d'Europe.* Clemont-Ferrand: Presses Universitaires Blaise Pascal, 2002.

Bertelli, Pino, and Maurizio Moretti, eds. *Una centrale al veleno: La battaglia del carbone a Piombino, 1968–1988.* Piombino: TraccEdizioni, 1989.

Berthold-Bond, Daniel. "The Ethics of Place: Reflections on Bioregionalism." *Environmental Ethics* 22, no. 1 (2000): 5–24.

Bess, Michael. *The Light Green Society: Ecology and Technological Modernity in France 1960–2000.* Chicago: University of Chicago Press, 2003.

Bettini, Maurizio. *Storia del porto di Livorno 1949–1994.* Livorno: Erasmo, 2004.

Bettini, Virginio. *Autoimbroglio ecologico e limiti allo sviluppo.* Florence: Sansoni, 1979.

Bevilacqua, Piero. "Scienza e ambiente: Quel che spetta al Novecento." *I Frutti di Demetra*, no. 16 (2008): 5–16.

Blanchetti, Emilia, Silvia Capotorto, and Emilio Conti. "Fenomenologia della sindrome nimby." *Limes: Rivista italiana di geopolitica*, no. 6 (2007). http://limes.espresso.repubblica.it/2007/11/22/il-clima-dellenergia-20/?p=359.

Bobbio, Luigi. "Democrazia e nuove forme di partecipazione." In *La democrazia in nove lezioni*, edited by Michelangelo Bovero and Valentina Pazé, 46–63. Roma-Bari: Laterza, 2010.

Bobbio, Luigi. "Il dibattito pubblico sulle grandi opere: Il caso dell'autostrada di Genova." *Rivista italiana di politiche pubbliche*, no. 1 (2010): 119–46.

Bocking, Stephen. *Nature's Experts: Science, Politics, and the Environment.* New Brunswick, NJ: Rutgers University Press, 2004.

Bondì, Roberto, and Antonello La Vergata. *Natura.* Bologna: il Mulino, 2015.

Bordignon, Fabio. *L'Europa unita . . . dall'antipolitica: Società, politica e partiti nell'Europa post-comunista.* Naples: Liguori, 2009.

Bortolotti, Lando, and Giuseppe De Luca. *Come nasce un'area metropolitana: Firenze, Prato, Pistoia, 1848–2000.* Florence: Alinea, 2000.

Bourdieu, Pierre. *La distinction: Critique sociale du jugement.* Paris: Éditions de Minuit, 1979.

Bovero, Michelangelo. "Democrazia al crepuscolo?" In *La democrazia in nove lezioni*, edited by Michelangelo Bovero and Valentina Pazé, 3–20. Roma-Bari: Laterza, 2010.

Boyer, Marc. *Il turismo: Dal gran tour ai viaggi organizzati.* Translated by Silvia Marzocchi. Turin: Universale Electa-Gallimard, 1997.

Branch, Michael P., ed. *Reading the Roots: American Nature Writing before Walden.* Athens: University of Georgia Press, 2004.

Brower, David R. *For Earth's Sake: The Life and Times of David Brower.* Salt Lake City, UT: Gibbs Smith Publisher, 1990.

Brown, Lester R. *Piano B 3.0: Mobilitarsi per salvare la civiltà.* Translated by Attilio Anitori. Milan: Edizioni Ambiente, 2008.

Buchanan, Colin. *Traffic in Towns: A Study of the Long Term Problems of Traffic in Urban Areas.* London: HMSO, 1963.

Bulgarelli, Vanni, and Catia Mazzeri. *La città e l'ambiente: Le trasformazioni ambientali urbane a Modena nel Novecento.* Modena: Apm Edizioni, 2009.

Bullard, Robert D., ed. *The Quest for Environmental Justice: Human Rights and the Politics of Pollution.* San Francisco: Sierra Club Books and University of California Press, 2005.

Burrell, George A. *Vitiation of Garage Air by Automobile Exhaust Gases.* Washington, DC: Bureau of Mines, 1919.

Cacciari, Massimo. "La pretesa faustiana dell'ecologismo." *Equilibri*, no. 3 (1998): 337–41.

Cacciari, Paolo, ed. *La società dei beni comuni: Una rassegna.* Rome: Ediesse, 2010.

Cadorna, Maria Clelia. *La storia della villeggiatura: Dall'epoca romana al Novecento.* Rome: Edizioni Abete, 1994.

Caglioti, Vincenzo. "L'uomo e l'ambiente." In Senato della Repubblica, *Problemi dell'ecologia.* Vol. 1, *Nota introduttiva del presidente Fanfani ed illustrazioni dei professori Caglioti, Marini-Bettolo, Nebbia, Passino, Pavan e Tomaselli.* Rome: Tipografia del Senato, 1971.

Calise, Mauro. *Il partito personale: I due corpi del leader.* Roma-Bari: Laterza, 2000.

Camera dei Deputati. *Indagini conoscitive e documentazioni legislative.* Vol. 7, *Indagine conoscitiva della XII Commissione permanente: Situazione e prospettive dell'industria automobilistica nazionale.* Rome, 1971.

Campos Venuti, Giuseppe. "Cinquant'anni: Tre generazioni urbanistiche." In *Cinquant'anni di urbanistica in Italia 1942–1992*, edited by Giuseppe Campos Venuti and Federico Oliva, 13–34. Roma-Bari: Laterza, 1993.

Campos Venuti, Giuseppe. *Città senza cultura: Intervista sull'urbanistica*. Roma-Bari: Laterza, 2010.

Campos Venuti, Giuseppe, and Federico Oliva, eds. *Cinquant'anni di urbanistica in Italia 1942–1992*. Roma-Bari: Laterza, 1993.

Campos Venuti, Giuseppe, and Odoardo Reali. "Firenze: L'urbanistica contrattata." In *Cinquant'anni di urbanistica in Italia 1942–1992*, edited by Giuseppe Campos Venuti and Federico Oliva, 313–27. Roma-Bari: Laterza, 1993.

Campus, Donatella. *L'antipolitica al governo: De Gaulle, Reagan, Berlusconi*. Bologna: il Mulino, 2006.

Cappelli, Andrea, and Silvano Simoni. "L'oro verde: Vizi e virtù dei biocarburanti; Il tempo che farà, le guerre dei tubi, l'Italia a rischio." *Limes: Rivista italiana di geopolitica*, no. 6 (2007): 145–57.

Caputo, Michele. "Un progetto per la salvaguardia del patrimonio ambientale italiano." *il Mulino* 56, no. 434 (2007): 1042–49.

Capuzzo, Paolo, ed. *Genere, generazione e consumi: L'Italia degli anni sessanta*. Rome: Carocci, 2003.

Carraro, Carlo, and Alessandra Mazzai. *Il clima che cambia: Non solo un problema ambientale*. Bologna: il Mulino, 2015.

Carriero, Libero. *Il costo di eliminazione dell'inquinamento in Italia*. Milan: FrancoAngeli, 1972.

Carson, Rachel. *Silent Spring*. Cambridge, MA: Riverside Press, 1962.

Castonguay, Stephane, and Matthew D. Evenden, eds. *Urban Rivers: Remaking Rivers, Cities, and Space in Europe and North America*. Pittsburgh: University of Pittsburgh Press, 2012.

Ceccarelli, Paolo, Bruno Gabrielli, and Renato Rozzi, eds. *Traffico urbano: Che fare?* Padua: Marsilio, 1968.

Cederna, Antonio. *I vandali in casa*. Roma-Bari: Laterza, 1956.

Cederna, Antonio. *Mirabilia Urbis: Cronache romane 1957–1965*. Turin: Einaudi, 1965.

Cederna, Antonio. *La distruzione*. Turin: Einaudi, 1975.

Celati, Giampiero, and Leo Gattini. *Sale e pietra*. Pisa: Giardini, 1993.

Celati, Giampiero, and Leo Gattini. *Cinquant'anni di vita*. Livorno: Nuova Fortezza, 1995.

CENSIS (Centro Studi Investimenti Sociali). *Energia, sviluppo, ambiente*. Milan: Garzanti, 1990.

Centemeri, Laura. *Ritorno a Seveso: Il danno ambientale, il suo riconoscimento, la sua riparazione*. Milan: Bruno Mondadori, 2006.

Centro Ornitologico Toscano. *Indagine bibliografica sull'impatto dei parchi eolici sull'avifauna: Relazione e bibliografia commentata.* Florence: Regione Toscana-Giunta Regionale, 2002.

Chiarugi, Enrico. "Legambiente: Slogan d'azione." *Equilibri* 1, no. 2 (1997): 287–91.

Chimenti, Anna. *Storia dei referendum: Dal divorzio alla riforma elettorale.* Milan: CDE, 1993.

Ciccone, Filippo. "Roma: Capitale senza piano." In *Cinquant'anni di urbanistica in Italia 1942–1992*, edited by Giuseppe Campos Venuti and Federico Oliva, 241–58. Roma-Bari: Laterza, 1993.

Ciccotti, Giovanni, and Marcello Cini. *L'ape e l'architetto: Paradigmi scientifici e materialismo storico.* Milan: Feltrinelli, 1976.

Cioc, Mark. *The Rhine: An Eco-Biography, 1815–2000.* Seattle: University of Washington Press, 2002.

Codeluppi, Vanni. *I consumatori: Storia, tendenze, modelli.* Milan: FrancoAngeli, 1992.

Colombi, A. *Seveso due anni dopo: Materiale di documentazione.* Milan: Unicopli, 1978.

Commissione Interministeriale per lo Studio della Sistemazione Idraulica e della Difesa del Suolo. *La sistemazione del Bacino dell'Arno: Relazione sul lavoro svolto al 31 dicembre 1968.* Rome, 1968.

Commissione Paritetica Comune di Piombino, ENEL per il Controllo della Centrale Termoelettrica di Torre del Sale. *Lo stato della qualità dell'aria.* Pisa: Tipografia pubblicazione editrice pisana, 1992.

Commissione per l'Indagine sul Traffico a Roma. *Il traffico a Roma: La situazione attuale e le previsioni fino al 1985.* Rome: Amministrazione comunale, 1968.

Commoner, Barry. *The Closing Circle: Nature, Man, and Technology.* New York: Knopf, 1971.

Comune di Firenze. *Firenzecologia.* Rome: Il Ventaglio, 1987.

Comune di Piombino. *L'ambiente Piombino: Cinque anni di attività per la tutela ambientale.* Piombino, 1991.

Confesercenti Regionale. *Il turismo in Toscana: Atti del Convegno.* Poggibonsi: Lalli, 1984.

Conti, Laura. *Che cos'è l'ecologia: Capitale, lavoro e ambiente.* Milan: G. Mazzotta, 1977.

Conti, Laura. *Visto da Seveso: L'evento straordinario e l'ordinaria amministrazione.* Milan: Feltrinelli, 1977.

Conti, Mario, ed. *L'impatto ambientale di centrali elettriche alimentate a biomasse legnose: Esperienze e valutazioni.* Rome: ENEA, 2003.

Corona, Gabriella. "Risorse nella città: Natura e territorio a Napoli tra Otto e

Novecento." In *Ambiente e risorse nel Mezzogiorno contemporane*, edited by Piero Bevilacqua and Gabriella Corona, 191–208. Rome: Donzelli, 2000.

Corona, Gabriella. "La sostenibilità urbana a Napoli: Caratteri strutturali e dinamiche storiche." *Meridiana: Rivista di storia e scienze sociali* 42 (2001): 15–43.

Corona, Gabriella. *I ragazzi del piano: Napoli e le ragioni dell'ambientalismo urbano*. Rome: Donzelli, 2007.

Corona, Gabriella. *A Short Environmental History of Italy: Variety and Vulnerability*. Winwick, UK: White Horse Press, 2017.

Corona, Gabriella, and Simone Neri Serneri. "Città e ambiente nell'Italia contemporanea: Un'introduzione." In *Storia e ambiente: Città, risorse e territori nell'Italia contemporanea*, edited by Gabriella Corona and Simone Neri Serneri, 11–37. Rome: Carocci, 2007.

Coumel, Laurent. "A Failed Environmental Turn? Khrushchev's Thaw and Nature Protection in Soviet Russia." *Soviet and Post-Soviet Review* 40, no. 2 (2013): 167–89.

Crainz, Guido. *Il paese reale: Dall'assassinio di Moro all'Italia di oggi*. Rome: Donzelli, 2012.

Cronon, William. *Nature's Metropolis: Chicago and the Great West*. New York: W. W. Norton, 1991.

Dahl, Robert A. *Polyarchy: Participation and Opposition*. New Haven, CT: Yale University Press, 1971.

Dahl, Robert A. *On Democracy*. New Haven, CT: Yale University Press, 1998.

Dalla Casa, Guido. *Ecologia profonda*. Turin: Pangea, 1996.

Daly, Herman E., ed. *Essays toward a Steady-State Economy*. Cuernavaca: Centre Intercultural de Documentación, 1971.

Damato, Mino E. *L'auto che non uccide*. Rome: L'editrice dell'automobile, 1969.

Danielis, Romeo. *I trasporti e l'ambiente: Aspetti di economia e politica dei trasporti*. Turin: Giappichelli, 1996.

Dauvergne, Peter. *The Shadows of Consumption: Consequences for the Global Environment*. Cambridge, MA: MIT Press, 2008.

Davis, H. Louise, Karyn Pilgrim, and Madhudaya Sinha, eds. *The Ecopolitics of Consumption: The Food Trade*. Lanham, MD: Lexington Books, 2016.

De Angelis, Giacomo. *Le acque dell'Arno*. Lanciano, 1969.

Dear, Michael. "Understanding and Overcoming the NIMBY syndrome." *Journal of the American Planning Association* 58, no. 3 (1992): 288–301.

Deléage, Jean Paul. *Storia dell'ecologia: Una scienza dell'uomo e della natura*. Translated by Tukery Capra. Naples: Cuen, 1991.

Dellapenna, Joseph W., and Joyeeta Gupta. *The Evolution of the Law and Politics of Water*. Dordrecht: Springer Netherlands, 2009.

Della Porta, Donatella. "La democrazia partecipativa e i movimenti sociali: Micro e macro dinamiche." In *Le nuove forme della partecipazione*, edited by Bianca R. Gelli. Rome: Carocci, 2007.

Della Porta, Donatella, and Mario Diani. *Movimenti senza protesta? L'ambientalismo in Italia.* Bologna: il Mulino, 2004.

Dell'Aquila, Anna. *Caserta (1945–1974): Una storia urbana e ambientale.* Arezzo: Fruska, 2013.

Della Seta, Roberto. *La difesa dell'ambiente in Italia: Storia e cultura del movimento ecologista.* Milan: FrancoAngeli, 2000.

Della Seta, Roberto, and Daniele Guastini. *Dizionario del pensiero ecologico: Da Pitagora ai no-global.* Rome: Carocci, 2007.

Della Valentina, Gianluigi. *Storia dell'ambientalismo in Italia: Lo sviluppo insostenibile.* Milan: Bruno Mondadori, 2011.

De Lucia, Vezio. *Se questa è una città: La condizione urbana nell'Italia contemporanea.* Rome: Donzelli, 2006.

Dematteis, Giuseppe. "Le trasformazioni territoriali e ambientali." In *Storia dell'Italia repubblicana.* Vol. 2, *La trasformazione dell'Italia: Sviluppo e squilibri.* Turin: Einaudi, 1995.

Descola, Philippe. "Le nature del mondo." *Equilibri*, no. 3 (2006): 469–87.

De Seta, Cesare. *Città, territorio e Mezzogiorno in Italia.* Turin: Einaudi, 1977.

De Steiguer, Joseph E. *The Origins of Modern Environmental Thought.* Tucson: University of Arizona Press, 2006.

Deverell, William, and Greg Hise, eds. *Land of Sunshine: An Environmental History of Metropolitan Los Angeles.* Pittsburgh: University of Pittsburgh Press, 2006.

Dewees, Donald N. *Automobile Air Pollution: An Economic Analysis.* Cambridge, MA: Harvard University Press, 1971.

Dewees, Donald N. *Economic and Public Policy: The Automobile Pollution Case.* Cambridge, MA: MIT Press, 1974.

Diamanti, Ilvo. "La democrazia degli interstizi: Società e partiti in Europa dopo la caduta del Muro." *Rassegna italiana di sociologia*, no. 3 (2007): 387–411.

Diamond, Jared. *Collapse: How Societies Choose to Fail or Succeed.* London: Penguin Books, 2006.

Diani, Mario. *Isole nell'arcipelago: Il movimento ecologista in Italia.* Bologna: il Mulino, 1988.

Diani, Mario, and Giovanni Lodi. "Three in One: Currents in the Milan Ecology Movement." *International Social Movements Research* 1 (1988): 103–24.

Dichter, Ernest. *Handbook of Consumer Motivations: The Psychology of the World of Objects.* New York: McGraw-Hill, 1964.

Dickie, John, John Foot, and Frank M. Snowden, eds. *Disastro! Disasters in Italy since 1860: Culture, Politics, Society.* New York: Palgrave, 2002.

Dittrich, Monika, Stefan Giljum, Stephan Lutter, and Christine Polzin. *Green*

Economics around the World? Implications of Resource Use for Development and the Environment. Vienna: Sustainable Europe Research Institute, 2012.

Donati, Paolo R. "Building a Unified Movement: Resource Mobilization, Media Work, and Organizational Transformation in the Italian Environmentalist Movement." *Research in Social Movements, Conflicts and Change* 19 (1996): 125–57.

Dowie, Mark. *Losing Ground: American Environmentalism at the End of the Twentieth Century,* Cambridge, MA: MIT Press, 1995.

Drengson, Alan, and Yuichi Inoue, eds. *The Deep Ecology Movement: An Introductory Anthology.* Berkeley, CA: North Atlantic Books, 1995.

Dumont, René. *Seule une écologie socialiste.* Paris: R. Laffont, 1977.

Durning, Alan Thein. *How Much Is Enough? The Consumer Society and the Future of the Earth.* New York: W. W. Norton, 1992.

Ehrlich, Paul R. *The Population Bomb.* New York: Ballantine Books, 1968.

EIA (Energy Information Administration). *International Energy Outlook 2008.* Washington, DC: EIA, 2008.

Emerson, Ralph W. *Nature and Other Essays.* Mineola, NY: Dover Publications, 2009.

ENEA (Comitato Nazionale per la Ricerca e lo Sviluppo dell'Energia Nucleare e delle Energie Alternative). *Libro Bianco per la valorizzazione delle fonti rinnovabili.* Rome: ENEA, 1999.

ENEA. *Energia eolica: Aspetti tecnici, ambientali e socio-economici.* Rome: ENEA, 2000.

ENEA. *Rapporto energia e ambiente 2005: I dati.* Rome: ENEA, 2005.

ENEA. *Rapporto energia e ambiente 2007: Analisi e scenari.* Rome: ENEA, 2008.

EPIA (European Photovoltaic Industry Association) and Greenpeace. *Solar Generation IV-2007: Solar Electricity for over One Billion People and Two Million Jobs by 2020.* Brussels: EPIA; Amsterdam: Greenpeace, 2007.

EPIA and Greenpeace. *Solar Generation V-2008: Solar Electricity for over One Billion People and Two Million Jobs by 2020.* Brussels: EPIA; Amsterdam: Greenpeace, 2008.

Esposito, John C., Ralph Nader, and Larry J. Silverman. *Vanishing Air: The Ralph Nader Study Group Report on Air Pollution.* New York: Grossman Publishers, 1970.

EurObserver. *Photovoltaic Energy Barometer 2006.* Paris, 2006.e

European Biomass Industry Association. *Annual Report.* Brussels, 2007.

European Commission. *Everything on Transport Statistics: Data 1970–2002* (dvd-rom). Luxemburg: European Communities, 2004.

European Commission. Directorate General for Energy and Transport. *Energy and Transport in Figures.* Bruxelles, 2003.

European Environmental Agency. *Urban Sprawl in Europe: The Ignored Challenge.* Copenhagen, 2006.

European Union. *Transport and Environment: Data 1980–1999.* Luxembourg, 2001.

European Union. *Energy and Transport in Figures*. Luxembourg, 2010.

European Wind Energy Association. *Wind Energy in Europe*. Brussels, 2007.

Eurostat. *Transport and Environment: Data 1980–1999*. Luxembourg, 2001.

Fabbri, Luciano. *La città inquinata*. Florence: Medicea, 1972.

Facco, Leonardo. *Elogio dell'antipolitica: In difesa delle libertà individuali*. Soveria Mannelli: Rubbettino, 2012.

Faith, William L. *Automobile Exhaust and Smog Formation*. San Marino, CA: Air Pollution Foundation, 1957.

Falco, Luigi. *Natura, monumenti, ambiente: La legislazione di tutela dal 1865 alle regioni*. Turin: Celid, 1979.

Fano, Pietro L., and Marco Ponti. *Il traffico urbano in Italia*. Milan: FrancoAngeli, 1972.

Fava, Valentina. "Production versus Mobility? New Perspectives for an Old Dilemma." In *Mobility in History: The State of the Art in the History of Transport, Traffic and Mobility*, edited by Gijs Mom, Gordon Pirie, and Laurent Tissot, 247–52. Neuchâtel: Éditions Alphil, 2009.

Felice, Emanuele. *Ascesa e declino: Storia economica d'Italia*. Bologna: il Mulino, 2015.

Ferrando, Stefania. "Una politica dell'avvenire: Conversazione con Bruno Latour." *Equilibri*, no. 2 (2015): 335–45.

Ferrari, Maria Luisa. "Trasformazioni urbane e igiene nell'industrializzazione ottocentesca di Verona, Vicenza e Padova." In *Storia e ambiente: Città, risorse e territori nell'Italia contemporanea*, edited by Gabriella Corona and Simone Neri Serneri, 55–85. Rome: Carocci, 2007.

Fieldner, Arno C. *Automobile Exhaust Gases in Vehicular Tunnels*. Washington, DC: Bureau of Mines, 1920.

Fishkin, James S. *La nostra voce: Opinione pubblica e democrazia, una proposta*. Venice: Marsilio, 2003.

Fitoussi, Jean-Paul, and Eloi Laurent. *La nouvelle écologie politique: Economie et développement humain*. Paris: Éditions du Seuil, 2008.

Flink, James J. *The Automobile Age*. Cambridge, MA: MIT Press, 1988.

Flower, Raymond. *One Hundred Years on the Road: A Social History of the Automobile*. New York: McGraw Hill, 1981.

Flynn, James R. *Senza alibi: Il cambiamento climatico; impedire la catastrofe*. Translated by Libero Sosio. Turin: Bollati Boringhieri, 2015.

Fondazione Giovanni Agnelli. *Volontari per l'arte e per l'ambiente*. Turin: Fondazione Giovanni Agnelli, 1985.

Foot, John. *Modern Italy*. Basingstoke, UK: Palgrave Macmillan, 2014.

Foster, Mark S. *A Nation on Wheels: The Automobile Culture in America since 1945*. Belmont, UK: Thomson, 2003.

Freudenberg, William R., and Susan K. Pastor. "NIMBYS and LULUS: Stalking the Syndromes." *Journal of Social Issues* 48 (1992): 39–61.

Fuga, Fabrizio. *L'industria del biossido di titanio: La tioxide di Scarlino, un case study.* Pisa: Pacini, 1996.

Fuhs, Allen E. *Hybrid Vehicles and the Future of Personal Transportation.* Boca Raton, FL: CRC Press, 2009.

Fukuyama, Francis. *The End of History and the Last Man.* New York: Free Press, 1992.

Fusaro, Sergio, and Francesco P. Tronca. *Le attività industriali a rischio di incidente rilevante: Analisi e commento della direttiva Seveso.* Padua: Cedam, 1990.

Gabellini, Patrizia. "Figure di urbanisti e programmi di urbanistica." In *Cinquant'anni di urbanistica in Italia 1942–1992*, edited by Giuseppe Campos Venuti and Federico Oliva, 440–67. Roma-Bari: Laterza, 1993.

Giardi, Dario, and Valeria Trapanese. *Dizionario dell'ambiente.* Florence: Alinea Editrice, 2007.

Gifford, Terry. *Reconnecting with John Muir: Essays in Post-Pastoral Practice.* Athens: University of Georgia Press, 2006.

Gigli, Franco. *Urbanistica di settore: Il traffico.* Rome: Multistampa, 1965.

Ginsborg, Paul. *Italy and Its Contents: Family, Civil Society, State, 1980–2001.* London: Penguin Press, 2001.

Giovannoni, Giulio. *Governare il territorio: Una riflessione sul caso toscano.* Milan: FrancoAngeli, 2004.

Giuntini, Andrea. "Nascita, sviluppo e tracollo della rete infrastrutturale." In *Storia d'Italia.* Vol. 15, *L'industria*, edited by Franco Amatori, Duccio Bigazzi, Renato Giannetti, and Luciano Segreto, 551–616. Turin: Einaudi, 1999.

Giuntini, Andrea, and Claudio Pavese, eds. *Reti, mobilità, trasporti: Il sistema italiano tra prospettiva storica e innovazione.* Milan: FrancoAngeli, 2004.

Givone, Silvia. "L'inceneritore della piana fiorentina: Una decisione senza partecipazione." In *Il bisogno di decidere*, edited by Massimo Morisi and Andrea Paci, 189–205. Bologna: il Mulino, 2009.

Goldsmith, Edward, and Robert Allen. *A Blueprint for Survival.* Harmondsworth, UK: Penguin, 1972.

Goria, Alessandra. "Istituzioni per l'uso: Unep." *Equilibri* 1, no. 1 (1997): 103–8.

Gorz, André. *Écologie et politique.* Paris: Galilée, 1975.

Grad, Frank P. *The Automobile and the Regulation of Its Impact on the Environment: A Study.* Norman: University of Oklahoma Press, 1975.

Greenpeace International. *How the Palm Oil Industry Is Cooking the Climate.* Amsterdam: Greenpeace, 2007.

Guarini, Elena Fasano, Giuseppe Petralia, and Paolo Pezzino. *Storia della Toscana.* Vol. 2, *Dal Settecento ad oggi.* Bari-Roma: Laterza, 2004.

Guarnieri, Gino. *Livorno marinara: Gli sviluppi portuali, la funzione economica, la tecnica commerciale marittima; Con tavole illustrative fuori testo e ricca appendice documentaria.* Livorno: Tip. Benvenuti e Cavaciocchi, 1962.

Guerrieri Ciacieri, Emanuele. *Bioregionalismo: La visione locale di un mondo globale.* Ragusa: Argo, 2006.

Guillerme, André, Anne-Cécile Lefort, and Gérard Jigaudon. *Dangereux, insalubries et incommodes: Paysages industriels en Banlieu parisienne XIX-XX Siècles.* Seyssel: Champ Vallon, 2004.

Hamilton, Clive. *Requiem for a Species: Why We Resist the Truth about Climate Change.* London: Earthscan, 2010.

Hardin, Garrett J. "The Tragedy of the Commons." *Science* 162, no. 3859 (1968): 1243–48.

Harrison, David. *Who Pays for Clean Air: The Cost and Benefit of Federal Automobile Emission Controls.* Cambridge, MA: Ballinger, 1975.

Hawkins, Gay, Emily Potter, and Kane Race. *Plastic Water: The Social and Material Life of Bottled Water.* Cambridge, MA: MIT Press, 2015.

Hays, Samuel P. *Conservation and the Gospel of Efficiency: The Progressive Conservation Movement 1890–1920.* Cambridge, MA: Harvard University Press, 1999.

Heinberg, Richard. *The End of Growth: Adapting to Our New Economic Reality.* Gabriola Island, BC: New Society Publishers, 2011.

Hill, Richard L. *Power from Wind: A History of Windmill Technology.* Cambridge: Cambridge University Press, 1994.

Hodge, Joseph M. *Triumph of the Expert: Agrarian Doctrines of Development and the Legacies of British Colonialism.* Athens: Ohio University Press, 2007.

Hofrichter, Richard, ed. *Toxic Struggles: The Theory and Practice of Environmental Justice.* Salt Lake City: University of Utah Press, 2002.

Hubbard, Phil. "Accommodating Otherness: Anti-asylum Centre Protest and the Maintenance of White Privilege." *Transactions of the Institute of British Geographers* 30, no. 1 (2005): 52–65.

Hughes, J. Donald, ed. *The Face of the Earth: Environment and World History.* Armonk, NY: M. E. Sharpe, 2000.

IEA (International Energy Agency). *World Energy Outlook 2006.* Paris: IEA Publications, 2006.

IEA. *World Energy Outlook 2008: Sinesti.* Paris: IEA Publications, 2008.

Illich, Ivan. *La convivialité.* Paris: Éditions du Seuil, 1973.

Il Mulino, 1951–2000: Indice storico. Bologna: il Mulino, 2001.

Indovina, Francesco, ed. *Lo spreco edilizio.* Padua: Marsilio, 1972.

Innocenti, Piero. *L'industria nell'area fiorentina.* Florence: Associazione degli industriali di Firenze, 1979.

IRPET (Istituto Regionale Programmazione Economica della Toscana). *Le zonizzazioni della Toscana 1754–1973.* Florence: Tip. Giuntina, 1988.

IRPET. *Il turismo in Toscana: Rapporto strutturale 1993–2000.* Florence: Regione Toscana, 2001.

ISFORT (Istituto Superiore di Formazione e Ricerca per i Trasporti). *Ottavo rapporto sulla mobilità urbana in Italia.* Rome, 2011.

Isherwood, Baron, and Mary Douglas. *World of Goods.* New York: Basic Books, 1979.

ISPRA (Istituto superiore per la protezione e la ricerca ambientale). *Annuario dei dati ambientali 2009.* Rome: ISPRA, 2009.

ISTAT (Istituto Nazionale di Statistica). *Statistiche ambientali: 1996.* Battipaglia: Emilcomp, 1995.

ISTAT. *Statistiche ambientali: 1998.* Pomezia: Albagraf, 1998.

ISTAT. *Flussi di materia dell'economia italiana: Conti anno 1997; Indicatori anni 1980–2001.* Rome, 2006.

ISTAT. *Statistiche ambientali 2007.* Rome: ISTAT, 2007.

Istituto per la documentazione e gli studi legislativi. *Inquinamenti dell'aria, dell'acqua e del suolo.* Milan: Giuffrè, 1971.

Istituto Superiore per la Protezione e la Ricerca Ambientale. *Annuario dei dati ambientali 2009.* Rome: ISPRA, 2009.

Italia Nostra. Ente Provinciale per il Turismo. *Inquinamento atmosferico: Processo all'automobile.* Milan: Grafiche A. Nava, 1970.

Jamison, Andrew. *The Steam-Powered Automobile: An Answer to Air Pollution.* Bloomington: Indiana University Press, 1970.

Jordan, Grant, and William A. Maloney. *The Protest Business? Mobilizing Campaign Groups.* Manchester: Manchester University Press, 1997.

Judt, Tony. *Ill Fares the Land.* New York: Penguin Press, 2010.

Judt, Tony. *L'età dell'oblio: Sulle rimozioni del '900.* Translated by Paolo Falcone. Roma-Bari: Laterza, 2011.

Juuti, Petri S., and Tapio S. Katko, eds. *From a Few to All: Long-Term Development of Water and Environmental Services in Finland.* Pieksämäki, Finland: KehräMedia, 2004.

King, Clyde L. *The Automobile: Its Province and Its Problems.* Philadelphia, 1924.

Klingle, Matthew. *Emerald City: An Environmental History of Seattle.* New Haven, CT: Yale University Press, 2007.

Kousis, Maria. "Economic Opportunities and Threats in Contentious Environmental Politics: A View from the European South." *Theory and Society* 33, no. 3–4 (2004): 393–415.

Ladd, Brian. *Autophobia: Love and Hate in the Automotive Age.* Chicago: University of Chicago Press, 2008.

Laird, Frank. *Technology Policy and Institutional Values.* Cambridge: Cambridge University Press, 2001.

La Mantia, Tommaso, Giuseppe Barbera, Rocco Lo Duca, Bruno Massa, and Salvatore Pasta. "Gli impatti degli impianti eolici sulla componente biotica e le misure di mitigazione." In *Eolico: Paesaggio e ambiente*, edited by Gianni Silvestrini and Mario Gamberale, 95–140. Rome: Franco Muzzio Editore, 2004.

Latouche, Serge. *L'occidentalizzazione del mondo: Saggio sul significato, la portata e i limiti dell'uniformazione planetaria.* Translated by Alfredo Salsano. Turin: Bollati Boringhieri, 1992.

Latouche, Serge. *Breve trattato sulla decrescita serena.* Translated by Fabrizio Grillenzoni. Turin: Bollati Boringhieri, 2007.

Latouche, Serge. *La scommessa della decrescita.* Translated by Matteo Schianchi. Milan: Feltrinelli, 2009.

Latouche, Serge. "Decrescita." In *La sinistra che verrà: Le parole chiave per cambiare*, edited by Giuliano Battiston and Giulio Marcon. Rome: Minimum fax, 2018.

Latour, Bruno. *Un monde pluriel mais commun: Entretiens avec François Ewald.* La Tour d'Aigues: Éditions de l'Aube, 2003.

Lattarulo, Patrizia, ed. *I costi ambientali e sociali della mobilità.* Milan: FrancoAngeli, 2003.

Legambiente. *Smog e dintorni 2006: L'inquinamento atmosferico e acustico nelle città italiane.* Rome, 2006.

Legambiente. *Comuni rinnovabili 2008: Rapporto di Legambiente.* Rome, 2008.

Legambiente. *Proposte per lo sviluppo del minieolico in Italia.* Rome, 2008.

Lehmann, E. J. *Automobile Air Pollution: A Bibliography with Abstracts.* Springfield: National Technical Information Service, 1973.

Leibbrand, Kurt. *Transportation and Town Planning.* Cambridge, MA: MIT Press, 1964.

Leone, Ugo. *L'Italia inquinata.* Naples: Esi, 1970.

Leopold, Aldo. "The Conservation Ethic." *Journal of Forestry* 31, no. 6 (1931): 634–43.

Leopold, Aldo. *A Sand County Almanac, and Sketches Here and There.* New York: Oxford University Press, 1949.

Lichtermann, Paul. *The Search for Political Community: American Activists Reinventing Tradition.* Cambridge: Cambridge University Press, 1996.

Löfstedt, Ragnar. "Sweden's Biomass Controversy: A Case Study of Communicating Policy Issues." *Environment* 40, no. 4 (1998): 42–45.

Logan, Michael F. *Desert Cities: The Environmental History of Phoenix and Tucson.* Pittsburgh: University of Pittsburgh Press, 2006.

Lopez, Gregore Pio, and Tara Laan. *Biofuels: At What Cost? Government Support for Biodiesel in Malaysia.* Geneva: Global Subsidies Initiative of the International Institute for Sustainable Development, 2008.

Lorenzini, Sara. "Ecologia a parole? L'Italia, l'ambientalismo globale e il rapporto ambiente-sviluppo intorno alla conferenza di Stoccolma." *Contemporanea*, no. 3 (2016): 395–418.

Lorenzoni, Gianni. "Lo sviluppo industriale di Prato." In *Storia di Prato*, 3:83–574. Prato: Cassa di Risparmi e Depositi, 1980.

Losacco, Ugo. "Notizie e considerazioni sulle inondazioni d'Arno in Firenze." *L'Universo*, no. 5 (1967): 720–820.

Lovelock, James. *Gaia: Nuove idee sull'ecologia*. Translated by Vania Bassan Landucci. Turin: Boringhieri, 1981.

Lovelock, James. *La rivolta di Gaia*. Translated by Massimo Scaglione. Milan: Rizzoli, 2006.

Luchetti, Rosella, and Graziella Poli. *La Magona di Piombino: 1944–1970*. Florence: La Nuova Italia, 1982.

Lungonelli, Michele. *La Magona d'Italia: Impresa, lavoro e tecnologie in un secolo di siderurgia toscana 1865–1975*. Bologna: il Mulino, 1991.

Luzzatto Fegiz, Pier Paolo. *Il volto sconosciuto dell'Italia: Dieci anni di sondaggi Doxa*. Milan: Giuffrè, 1956.

Luzzi, Saverio. *Il virus del benessere: Ambiente, salute, sviluppo nell'Italia repubblicana*. Roma-Bari: Laterza, 2009.

Luzzi, Saverio. "Come l'Italia (non) uscì dal nucleare." *I Frutti di Demetra* 23 (2010): 45–65.

MacDonald, Austin F. *Planning for City Traffic*. Philadelphia, 1927.

Madrigal, Alexis. *Powering the Dream: The History and Promise of Green Technology*. Cambridge, MA: Da Capo Press, 2011.

Maggi, Stefano. *Storia dei trasporti in Italia*. Bologna: il Mulino, 2005.

Majocchi, Donatella, and Dadi Sanna, eds. *Inquinamento atmosferico: Processo all'automobile*. Milan: Grafiche A. Nava. Italia Nostra-Csap-Ept, 1970.

Mammarella, Luigi. *L'inquinamento in Italia*. Rome: Isvet, 1970.

Mariotti, Giuliana. *Storia del turismo*. Rome: Edizioni Saturna, 1958.

Marshall, Peter H. *Nature's Web: An Exploration of Ecological Thinking*. London: Simon & Schuster, 1992.

Massard-Guilbaud, Geneviève, and Richard Rodger, eds. *Environmental and Social Justice in the City: Historical Perspectives*. Cambridge: White Horse Press, 2011.

Mattei, Ugo. *Beni comuni: Un manifesto*. Roma-Bari: Laterza, 2011.

Mazzoni, Marco, ed. *Rapporto sullo stato delle acque dei principali fiumi in Toscana*. Florence: Centro stampa 2P, 2001.

Mazzoni, Marco, ed. *Rapporto sullo stato delle acque marine in Toscana*. Florence: Centro stampa 2P, 2001.

Mcavoy, Gregory E. "Partisan Probing and Democratic Decision Making: Rethinking the NIMBY Syndrome." *Policy Studies Journal* 26, no. 2 (1998): 274–92.

McCarthy, Tom. *Auto Mania: Cars, Consumers and Environment.* London: Yale University Press, 2007.

McCarthy, Tom. "Automobili, politica ambientale e individualismo liberale: Un dilemma americano." *I Frutti di Demetra* 21 (2009): 15–32.

McCarthy, Tom. "A Natural Intersection: A Survey of Historical Work on Mobility and the Environment." In *Mobility in History: The State of the Art in the History of Transport, Traffic and Mobility,* edited by Gijs Mom, Gordon Pirie, and Laurent Tissot, 61–81. Neuchâtel: Editions Alphil-Presses universitaires suisses, 2009.

McNeill, John R. *Qualcosa di nuovo sotto il sole: Storia dell'ambiente nel XX secolo.* Translated by Piero Arlorio. Turin: Einaudi, 2002.

Meadows, Donella H., and Dennis L. Meadows. *The Limits to Growth: A Report for the Club of Rome's Project on the Predicament of Mankind.* New York: Universe Books, 1972.

Melosi, Martin V. *The Sanitary City: Urban Infrastructure in America from Colonial Times to the Present.* Baltimore: Johns Hopkins University Press, 2000.

Melosi, Martin V. *Effluent America: Cities, Industry, Energy and the Environment.* Pittsburgh: University of Pittsburgh Press, 2001.

Melosi, Martin V., and Joseph A. Pratt, eds. *Energy Metropolis: An Environmental History of Houston and the Gulf Coast.* Pittsburgh: University of Pittsburgh Press, 2007.

Menna, Pietro. *L'energia pulita: Sole, acqua, vento . . . le fonti di energia rinnovabili.* Bologna: il Mulino, 2003.

Mercanti, Alberto, Rosanna Granatella, and Alessandra La Manna. *Energia eolica: Tra passato e futuro un'alternativa attuale.* Florence: Alinea, 2002.

Meyer, Edgar H. *I pionieri dell'ambiente: L'avventura del movimento ecologista italiano.* Milan: Carabà, 1995.

Midulla, Mariagrazia. "WWF: Prima i fatti poi le parole." *Equilibri* 1, no. 3 (1997): 431–37.

Mikler, John. *Greening the Car Industry: Varieties of Capitalism and Climate Change.* Cheltenham, UK: Edward Elgar, 2009.

Miller, Char. *Gifford Pinchot and the Making of Modern Environmentalism.* Washington, DC: Island Press, 2001.

Miller, Char, ed. *Cities and Nature in the American West.* Reno: University of Nevada Press, 2010.

Miller, Clark A., and Paul N. Edwards, eds. *Changing the Atmosphere: Expert Knowledge and Environmental Governance.* Cambridge, MA: MIT Press, 2001.

Ministero delle Infrastrutture e dei Trasporti. *Conto nazionale delle infrastrutture e dei trasporti anni 2008–2009.* Rome, 2010.

Ministry of Transportation. *National Account of Infrastructures and Transport.* Rome: Istituto Poligrafico e Zecca dello Stato, 2002.

Mishan, Ezra J. *The Costs of Economic Growth*. Harmondsworth, UK: Penguin, 1969.

Mitchell, Robert B., and Chester Rapkin. *Urban Traffic: A Function of Land Use*. New York: Columbia University Press, 1954.

Mitchell, William J., Christopher E. Borroni-Bird, and Lawrence D. Burns. *Reinventing the Automobile: Personal Urban Mobility for the 21st Century*. Cambridge, MA: MIT Press, 2010.

Mom, Gijs. *The Electric Vehicle: Technology and Expectations in the Automobile Age*. Baltimore: Johns Hopkins University Press, 2004.

Mom, Gijs. *Atlantic Automobilism: Emergence and Persistence of the Car 1895–1940*. New York: Berghahn, 2014.

Moraglio, Massimo. "Niche Development or Wider Renaissance? Italian Mobility History Growth in the Last Decade." In *Mobility in History: The State of the Art in the History of Transport, Traffic and Mobility*, edited by Gijs Mom, Gordon Pirie, and Laurent Tissot, 167–72. Neuchâtel: Editions Alphil-Presses universitaires suisses, 2009.

Moraglio, Massimo. "Una forma goffa e sgraziata: Paesaggio e autostrade in Italia nel secondo dopoguerra." *I Frutti di Demetra* 21 (2009): 33–46.

Mori, Giorgio. "Dall'unità alla guerra: Aggregazione e disgregazione di un'area regionale." In *Storia d'Italia: Le Regioni dall'unità a oggi, La Toscana*, edited by Giorgio Mori. Turin: Einaudi, 1986.

Mori, Giorgio, ed. *Storia d'Italia: Le Regioni dall'unità a oggi, La Toscana*. Turin: Einaudi, 1986.

Morin, Edgar. *Le Paradigme perdu, la nature humaine*. Paris: Éditions du Seuil, 1973.

Morisi, Massimo, and Annick Magnier. *Governo del territorio: Il modello Toscana*. Bologna: il Mulino, 2003.

Moscovici, Serge. *La Société contre nature*. Paris: Union générale d'éditions, 1972.

Mosley, Stephen. *The Chimney of the World: A History of Smoke Pollution in Victorian and Edwardian Manchester*. Cambridge: White Horse Press, 2001.

Mossa Verre, Marcello, ed. *Analisi del rischio per l'area di Piombino e strategie d'intervento: Piano di risanamento per le aree critiche a elevata concentrazione industriale di Livorno e Piombino*. Florence: ARPAT, 2000.

Muir, John. *Nature Writings: The Story of My Boyhood and Youth, My First Summer in the Sierra, the Mountains of California, Stickeen; Selected Essays*. New York: Penguin Books, 1997.

Mukerji, Chandra. *From Graven Images: Patterns of Modern Materialism*. New York: Columbia University Press, 1983.

Nader, Ralph. *L'auto che uccide*. Translated by G. Boschi Breghelli. Milan: Bompiani, 1967.

Naess, Arne. *Ecology, Community, and Lifestyle: Outline of an Ecosophy*. Cambridge: Cambridge University Press, 1989.

Nebbia, Giorgio. *Scritti di storia dell'ambiente e dell'ambientalismo 1970–2013*. Edited by Luigi Piccioni. Brescia: Fondazione Luigi Micheletti-Valori, 2004.

Nebbia, Giorgio, and Luigi Piccioni. "Un tornante del dibattito italiano sull'ambiente: La ricezione dei Limiti dello sviluppo." *Ricerche Storiche* 41, no. 3 (2011): 519–40.

Nencini, Franco. *Firenze, i giorni del diluvio*. Florence: Sansoni, 1966.

Neri, Valerio. "Fai, modelli organizzativi e dialettica istituzionale di un'associazione post-ambientalista." *Equilibri*, no. 2 (2004): 271–77.

Neri Serneri, Simone. "Uso e abuso delle risorse: L'emergere della questione ambientale." In *Storia del territorio e storia dell'ambiente: La Toscana contemporanea*, edited by Simone Neri Serneri, 101–16. Milan: FrancoAngeli, 2002.

Neri Serneri, Simone. *Incorporare la natura: Storie ambientali del Novecento*. Rome: Carocci, 2005.

Newman, Lance. *Our Common Dwelling: Henry Thoreau, Transcendentalism, and the Class Politics of Nature*. New York: Palgrave Macmillan, 2005.

Norton, Peter, Gijs Mom, Liz Millward, and Mathieu Flonneau, eds. *Mobility in History: Reviews and reflections*. Neuchâtel: Éditions Alphil-Presses universitaires suisses, 2011.

OECD (Organization for Economic Co-operation and Development). *Environmental Data: Compendium 2004*. Paris: OECD Publishing, 2004.

Officina delle idee di Rete@Sinistra. "La società dei beni comuni." In *La società dei beni comuni: Una rassegna*, edited by Paolo Cacciari, 17–26. Rome: Ediesse, 2010.

Ogle, Maureen. *In Meat We Trust: An Unexpected History of Carnivore America*. Boston: Houghton Mifflin Harcourt, 2013.

Oliva, Federico. "Milano: La 'deregulation' urbanistica." In *Cinquant'anni di urbanistica in Italia 1942–1992*, edited by Giuseppe Campos Venuti and Federico Oliva, 223–40. Roma-Bari: Laterza, 1993.

Olivito, Elisa. "Partecipazione e ambiente: Una sguardo critico." *Diritto Pubblico*, no. 2 (2011): 559–610.

OPEC (Organization of the Petroleum Exporting Countries). *World Oil Outlook 2007*. Vienna: OPEC Secretariat, 2007.

Opie, John. *Ogallala: Water for a Dry Land*. Lincoln: University of Nebraska Press, 2000.

Osti, Giorgio. "Come inquadrare i conflitti ambientali." In *Ambiente Italia 2007: Rapporto annuale di Legambiente*, edited by Duccio Bianchi and Giuseppe Gamba, 15–31. Milan: Edizioni Ambiente, 2007.

Ottati, Davis. *L'acquedotto di Firenze dal 1860 ad oggi*. Florence: Vallecchi, 1983.

Ottati, Davis. *Il ventre di Firenze: Storia della fognatura dall'epoca romana a oggi*. Florence: Editoriale Olimpia, 1999.

Owens, Susan. "Experts and the Environment: The UK Royal Commission on Environmental Pollution 1970–2011." *Journal of Environmental Law* 24, no. 1 (2012): 1–22.

Paccino, Dario. *L'imbroglio ecologico: L'ideologia della natura.* Turin: Einaudi, 1972.

Pallante, Maurizio. *La decrescita felice: La qualità della vita non dipende dal Pil.* Rome: Editori Riuniti, 2005.

Panciatici, Pier Luigi. *La siderurgia a Piombino: Uno stabilimento e una città.* Piombino: Officina libraria, 1996.

Paolini, Federico. "La più grossa sciagura sarebbe se la Fiat entrasse in crisi: Gianni Agnelli, le automobili e l'Italia nei tardi anni Sessanta." *Contemporanea*, no. 1 (Jan. 2004): 85–96.

Paolini, Federico. *Un paese a quattro ruote: Automobili e società in Italia.* Venice: Marsilio, 2005.

Paolini, Federico. *Storia sociale dell'automobile in Italia.* Rome: Carocci, 2007.

Paolini, Federico. "Automobili e ambiente: Verso una storia ambientale della mobilità?" *I Frutti di Demetra* 21 (2009): 5–13.

Paolini, Federico. *Breve storia dell'ambiente nel Novecento.* Rome: Carocci, 2009.

Paolini, Federico. "A Country on Four Wheels: The Car and Society in Italy (1900–1974)." *TsT- Transportes, Servicios y Telecomunicaciones* 17 (2009): 108–30.

Paolini, Federico. "Greening Mobility History in Italy: Toward an Interdisciplinary Way to Environmental Mobility Studies?" In *Mobility in History: Reviews and Reflections*, edited by Peter Norton, Gijs Mom, Liz Millward, and Mathieu Flonneau, 145–50. Neuchâtel: Éditions Alphil-Presses universitaires suisses, 2011.

Paolini, Federico. *Firenze 1946–2005: Una storia urbana e ambientale.* Milan: FrancoAngeli, 2014.

Parisi, Roberto. "Verso una città salubre: Lo spazio produttivo a Napoli tra storia e progetto." *Meridiana* 42 (2001): 53–74.

Parisi, Roberto. "Le forme d'Igea e le persuasioni di Prometeo: Fabbriche, sanatori e città giardino a Napoli." In *Storia e ambiente: Città, risorse e territori nell'Italia contemporanea*, edited by Gabriella Corona and Simone Neri Serneri, 123–40. Rome: Carocci, 2007.

Parrinello, Giacomo. *Fault Lines: Earthquakes and Urbanism in Modern Italy.* Oxford: Berghahn Books, 2015.

Pellizzoni, Luigi, ed. *Conflitti ambiental: Esperti, politica, istituzioni nelle controversie ecologiche.* Bologna: il Mulino, 2011.

Pellow, David N. "Framing Emerging Environmental Movement Tactics: Mobilizing Consensus, Demobilizing Conflict." *Sociological Forum* 14, no. 4 (1999): 659–83.

Pellow, David N., and Robert J. Brulle, eds. *Power, Justice, and the Environment: A Critical Appraisal of the Environmental Justice Movement.* Cambridge, MA: MIT Press, 2006.

Perlin, John. *From Space to Earth: The Story of Solar Electricity.* Ann Arbor, MI: Aatec Publications, 1999.

Piccioni, Luigi. *Primo di cordata: Renzo Videsott dal sesto grado alla protezione della natura.* Trento: Tipografia Editrice Temi, 2010.

Piccioni, Luigi. *Il volto amato della Patria: Il primo movimento per la protezione della natura in Italia 1880–1934.* Trento: Temi, 2014.

Piccioni, Luigi. "Nazione, patrimonio, paesaggio: Alle origini del moderno ambientalismo in Europa 1865–1914." *Storia e Futuro,* no. 38 (June 2015). http://storiaefuturo.eu/nazione-patrimonio-paesaggio-alle-origini-del-moderno-ambientalismo-in-europa-1865-1914/.

Pigou, Arthur C. *The Economics of Welfare.* London: Macmillan, 1920.

Piketty, Thomas. *L'économie des inégalités.* Paris: La Découverte & Syros, 1997.

Piketty, Thomas. *Le capital au XXI siècle.* Paris: Éditions du Seuil, 2013.

Piller, Charles. *The Fail-Safe Society: Community Defiance at the End of American Technological Optimism.* New York: Basic Books, 1991.

Pinchot, Gifford. *Breaking New Ground.* Washington, DC: Island Press, 1998.

Platt, Harold L. *Shock Cities: The Environmental Transformation and Reform of Manchester and Chicago.* Chicago: University of Chicago Press, 2005.

Poggio, Andrea. *Ambientalismo.* Milan: Editrice Bibliografica, 1996.

Pozzana, Giuseppe. *Livorno e la sua provincia: Materiali per una storia dello sviluppo economico e sociale dall'unità d'Italia ai giorni nostri; Rassegna bibliografico-critica.* Milan: FrancoAngeli, 1987.

Preti, Domenico. "Tra crisi e dirigismo: L'economia toscana del periodo fascista." In *Storia d'Italia: Le Regioni dall'unità a oggi, La Toscana,* edited by Georgio Mori. Turin: Einaudi, 1986.

Princen, Thomas, Michael Maniates, and Ken Conca, eds. *Confronting Consumption.* Cambridge, MA: MIT Press, 2002.

Principe, Ilario, and Paolo Sica. "L'inondazione di Firenze del 4 novembre 1966." *L'Universo* 47, no. 2 (1967): 192–222.

Provincia di Firenze. *La Provincia di Firenze e l'Unione regionale delle provincie toscane contro le calamità naturali per una politica di difesa del suolo e di riordino idrogeologico: Note e documenti 1956–1966.* Florence, 1966.

Rajan, Raghuram G. *Fault Lines: How Hidden Fractures Still Threaten the World Economy.* Princeton, NJ: Princeton University Press, 2010.

Ramírez Pérez, Sigfrido. "Un'auto più pulita e sicura: La Comunità europea e l'avvio della standardizzazione per gli autoveicoli (1952–1970)." *I Frutti di Demetra,* no. 21 (July 2010): 47–61.

Ramsay, William. *Bioenergy and Economic Development: Planning for Biomass Energy Programs in the Third World.* Boulder, CO: Westview Press, 1985.

Regione Toscana. *Mappa degli inquinamenti idrici della Toscana.* Florence, 1974.

Regione Toscana. *Progetto pilota per la sistemazione del bacino dell'Arno: Rapporto finale.* Vol. 2, *Organizzazione del progetto.* Florence: Regione Toscana, 1978.

Regione Toscana. *Piano regionale di risanamento delle acque: Primo programma di intervento 1980–1982.* Florence, 1980.

Regione Toscana. *Piano Regionale di Azione Ambientale 2004–2006.* Florence: Edifir, 2007.

Regione Toscana. *Rapporto sulle politiche agricole e rurali, allegato 1, cap. 3.1: Analisi della situazione in termini di punti di forza e di debolezza.* Supplement to *Bollettino ufficiale della Regione Toscana,* no. 48 (Nov. 28, 2007).

Regione Toscana. *Piano di indirizzo energetico regionale (PIER): Proposta della Giunta Regionale.* Florence, 2008.

Regione Toscana. *La qualità delle acque marine costiere in Toscana.* Florence, 2010.

Reuss, Martin. "Floods." In *Encyclopedia of World Environmental History.* Vol. 2, edited by Shepard Krech, John R. McNeill, and Carolyn Merchant, 535–41. New York: Routledge, 2004.

Ridker, Ronald G. *Economic Costs of Air Pollution: Studies in Measurement.* New York: F. A. Praeger, 1967.

Rocca, Lorena. "Partecipazione come pratica territoriale." *Equilibri,* no. 1 (2007): 117–24.

Roccato, Michele, and Terri Mannarini. *Non nel mio giardino: Prendere sul serio i movimenti NIMBY.* Bologna: il Mulino, 2012.

Rootes, Christopher, ed. *Environmental Movements: Local, National and Global.* London: Frank Cass, 1999.

Rootes, Christopher. "The Transformation of Environmental Activism: Activists, Organizations and Policy-Making." *Innovation: The European Journal of Social Sciences* 12 (1999): 153–73.

Ruzzenenti, Marino. *Un secolo di cloro e . . . PCB: Storia delle industrie Caffaro di Brescia.* Milan: Jaca Book, 2001.

Salmon, Christian. *La politica nell'era dello storytelling.* Translated by Nicola Vincenzoni. Rome: Fazi Editore, 2014.

Salvadori, Massimo L. *Democrazie senza democrazia.* Roma-Bari: Laterza, 2009.

Salzano, Edoardo. "L'Istituto nazionale di urbanistica." In *Cinquant'anni di urbanistica in Italia 1942–1992,* edited by Giuseppe Campos Venuti and Federico Oliva, 424–39. Roma-Bari: Laterza, 1993.

Sandquist, Karin. *The Appeal of Automobiles: Human Desires and the Proliferation of Cars.* Stockholm: Swedish Transport and Communications Research Board, 1997.

Sartori, Giovanni. *Homo videns: Televisione e post-pensiero.* Roma-Bari: Laterza, 1997.

Saviane, Giorgio, and Mario Cresci. *Voi di Rosignano.* Milan: Solvay, 1988.

Sawin, Janet L., and Kristen Hughes. "Dare energia alle città." In Worldwatch Institute, *State of the World 2007: Il nostro futuro urbanizzato*, 228–35. Milan: Edizioni Ambiente, 2007.

Sayers, Royd R. *Effect of Repeated Daily Exposure of Several Hours to Small Amounts of Automobile Exhaust Gas.* Washington, DC: Bureau of Mines, 1929.

Scattoni, Paolo. *L'urbanistica dell'Italia contemporanea: Dall'unità ai giorni nostri.* Milan: Newton & Compton, 2004.

Scheer, Hermann. *Il solare e l'economia globale: Energia rinnovabile per un futuro sostenibile.* Translated by Margherita Romaro. Milan: Edizioni Ambiente, 2004.

Schively, Carissa. "Understanding the NIMBY and LULU Phenomena: Reassessing Our Knowledge Base and Informing Future Research." *Journal of Planning Literature* 21, no. 3 (2007): 255–66.

Schlosberg, David. *Environmental Justice and the New Pluralism: The Challenge of Difference for Environmentalism.* New York: Oxford University Press, 1999.

Schott, Dieter, Bill Luckin, and Geneviève Massard-Guilbaud, eds. *Resources of the City: Contributions to an Environmental History of Modern Europe.* Aldershot, UK: Ashgate, 2005.

Schott, Dieter, and Michael Toyka-Seid, eds. *Die europäische Stadt und ihre Umwelt.* Darmstadt: WBG, 2008.

Schramm, Engelbert. "Experts in the Smelter Smoke Debate." In *The Silent Countdown: Essays in European Environmental History*, edited by Peter Brimblecombe and Christian Pfister, 196–209. New York: Springer-Verlag, 1990.

Schumacher, Ernst F. *Small Is Beautiful: A Study of Economics as if People Mattered.* London: Blond and Briggs, 1973.

Scichilone, Laura. *L'Europa e la sfida ecologica: Storia della politica ambientale europea (1969–1998).* Bologna: il Mulino, 2008.

Secchi, Bernardo. "Il settore edilizio e fondiario in un processo di sviluppo economico." In *Lo spreco edilizio*, edited by Francesco Indovina, 3–46. Padua: Marsilio, 1972.

Senato della Repubblica. *Problemi dell'ecologia.* 2 vols. Rome, 1971.

Shapiro, Judith. *China's Environmental Challenges.* Malden, MA: Polity Press, 2012.

Sica, Paolo. *Traffico e forma urbana.* Florence: Coop. Libraria Usf, 1967.

Sievert, James. *The Origins of Nature Conservation in Italy.* New York: Peter Lang, 2000.

Smil, Vaclav, and Kazuhiko Kobayashi. *Japan's Dietary Transition and Its Impacts.* Cambridge, MA: MIT Press, 2012.

Smith, Carl. *City Water, City Life: Water and the Infrastructure of Ideas in Urbanizing Philadelphia, Boston, and Chicago.* Chicago: University of Chicago Press, 2013.

Smith, Eric, and Marquez, Marisela. "The Other Side of the NIMBY Syndrome." *Society and Natural Resources* 13, no. 3 (2000): 273–80.

Solvay S.A. *Solvay: 125° anniversario 1863–1988*. Bruxelles: Solvay, 1988.

Squillante, Renato. "Traffico e ambiente." In ACI, *XXX Conferenza del traffico e della circolazione di Stresa*. Rome, 1973.

Steinberg, Theodore. *Nature Incorporated. Industrialization and the Waters of New England*. Cambridge: Cambridge University Press, 1991.

Stern, Nicholas. *A Blueprint for a Safer Planet: How to Manage Climate Change and Create a New Era of Progress and Prosperity*. London: Bodley Head, 2009.

Stradling, David. *Smokestacks and Progressives: Environmentalists, Engineers and Air Quality in America 1881–1951*. Baltimore: Johns Hopkins University Press, 1999.

Stradling, David, ed. *Conservation in the Progressive Era: Classic Texts*. Seattle: University of Washington Press, 2004.

Stradling, David. *The Nature of New York: An Environmental History of the Empire State*. Ithaca, NY: Cornell University Press, 2010.

Sullo, Fiorentino. *Lo scandalo urbanistico: Storia di un progetto di legge*. Florence: Vallecchi, 1964.

Szasz, Andrew. *Eco-populism: Toxic Waste and Movements for Environmental Justice*. London: UCL Press, 1994.

Tarr, Joel A. "Water and Wastes: A Retrospective Assessment of Wastewater Technology in the United States 1800–1932." *Technology and Culture* 25 (Apr. 1984): 226–63.

Tarr, Joel A. *The Search for the Ultimate Sink: Urban Pollution in Historical Perspective*. Akron: University of Ohio Press, 1996.

Tarr, Joel A., ed. *Devastation and Renewal: An Environmental History of Pittsburgh and Its Region*. Pittsburgh: University of Pittsburgh Press, 2003.

Tecneco. *Prima relazione sulla situazione ambientale del Paese*. 3 vols. Rome: Tip. Carlo Colombo, 1973.

Testa, Chicco, and Patrizia Feletig. *Contro (la) natura: Perché la natura non è buona, né giusta, né bella*. Venice: Marsilio, 2014.

Thomas, Keith. *L'uomo e la natura: Dallo sfruttamento all'estetica dell'ambiente, 1500–1800*. Translated by Elda Negri Monateri. Turin: Einaudi, 1994.

Thoms, David, Len Holden, and Tim Claydon. *The Motor Car and Popular Culture in the 20th Century*. Aldershot, UK: Ashgate, 1998.

Thomson, Michael J. *Great Cities and Their Traffic*. London: Gollancz, 1977.

Thoreau, Henry D. *Walden (Life in the Woods)/Civil Disobedience*. New York: Signet Classics, 2004.

Tinacci Mossello, Maria. *L'organizzazione del territorio pratese durante la seconda guerra mondiale e la ricostruzione*. Florence: Università degli Studi, 1983.

Treves, Liliana, ed. *Le risorse economiche del turismo in Toscana*. Padua: Cedam, 1967.

Tripp, Herbert A. *Town Planning and Road Traffic*. London: E. Arnold and Co., 1942.

Tuccari, Francesco. "Democrazie acefale e dispotismo postdemocratico." *Storia del pensiero politico* 1, no. 1 (2012): 105–41.

Uekoetter, Frank. *The Age of Smoke: Environmental Policy in Germany and the United States, 1880–1970.* Pittsburgh: University of Pittsburgh Press, 2009.

Unione Petrolifera. *Previsioni di domanda energetica e petrolifera italiana 2007–2020.* Rome: U.P., 2007.

van der Heijden, Hein-Anton. "Political Opportunity Structure and the Institutionalization of the Environmental Movement." *Environmental Politics* 6, no. 4 (1997): 25–50.

Ventura, Nico. *Città e autostrade negli squilibri regionali.* Padua: Patron, 1978.

Vercelli, Alessandro, and Simone Borghesi. *La sostenibilità dello sviluppo globale.* Rome: Carocci, 2005.

Vitale, Ermanno. *Contro i beni comuni: Una critica illuminista.* Roma-Bari: Laterza, 2013.

Wells, Christopher W. "La morte del Modello T: Strade pavimentate, auto coperte e tecnologia desueta." *I Frutti di Demetra* 21 (2009): 63–75.

Wells, Christopher W. *Car Country: An Environmental History.* Seattle: University of Washington Press, 2013.

WHO (World Health Organization). *Health Impact Assessment of Air Pollution in the Eight Major Italian Cities.* Bonn: WHO European Centre for Environment and Health, 2002.

Wolsink, Maarten. "Wind Power and the NIMBY-Myth: Institutional Capacity and the Limited Significance of Public Support." *Renewable Energy* 21, no. 1 (2000): 49–64.

World Resources Institute. *Greenhouse Gas Data and International Climate Policy.* Washington, DC: World Resources Institute, 2005.

World Resources Institute. *Navigating the Numbers.* Washington, DC: World Resources Institute, 2005.

World Resources Institute. *World Resources 2005: The Wealth for the Poor; Managing Ecosystems to Fight Poverty.* Washington, DC: World Resources Institute, 2005.

Worldwatch Institute. *State of the World 2004: Special Focus, the Consumer Society.* Worldwatch Institute, 2004.

Worldwatch Institute. *Biofuels for Transport: Global Potential and Implications for Sustainable Agriculture and Energy in the 21th Century.* London: Earthscan, 2007.

Worster, Donald. *Rivers of Empire: Water, Aridity and the American West.* Oxford: Oxford University Press, 1985.

Worster, Donald. *Storia delle idee ecologiche.* Translated by Elisa Gunella. Bologna: il Mulino, 1994.

Worster, Donald. *A Passion for Nature: The Life of John Muir.* New York: Oxford University Press, 2008.

Zambrini, Mario. "La valutazione degli impatti ambientali degli impianti eolici." In *Ambiente Italia 2007: Rapporto annuale di Legambiente*, edited by Duccio Bianchi and Giuseppe Gamba, 165–77. Milan: Edizioni Ambiente, 2007.

Zanardi, Nicola. "Greenpeace: Obiettivo mass media." *Equilibri* 1, no. 1 (1997): 117–21.

Ziglioli, Bruno. *La mina vagante: Il disastro di Seveso e la solidarietà nazionale.* Milan: FrancoAngeli, 2010.

Ziglioli, Bruno. *Sembrava nevicasse: La Eternit di Casale Monferrato e la Fibronit di Broni; Due comunità di fronte all'amianto.* Milan: FrancoAngeli, 2016.

Žižek, Slavoj. *Welcome to Interesting Times!* London: Verso Books, 2010.

Zolo, Danilo. *Il principato democratico.* Milan: Feltrinelli, 1992.

Zoppi, Mariella. *Firenze e l'urbanistica: La ricerca del piano; La politica amministrativa e i piani urbanistici dalla ricostruzione ad oggi.* Florence: Edizioni delle autonomie, 1982.

Zucconi, Guido. *La città contesa: Dagli ingegneri sanitari agli urbanisti (1885–1942).* Milan: Jaca Book, 1989.

INDEX

Note: Page numbers in *italics* refer to figures.